AF862492

SEAWATCHERS

Oceanographers in Action

SEAWATCHERS

Oceanographers in Action

BY

WILLIAM BIXBY

Illustrated by JOHN FLYNN

DAVID McKAY COMPANY, INC.

NEW YORK

1967

FORT WORTH PUBLIC LIBRARY

SEAWATCHERS
OCEANOGRAPHERS IN ACTION

COPYRIGHT © 1967 BY
WILLIAM BIXBY

All rights reserved, including the right to reproduce this book, or parts thereof, in any form, except for the inclusion of brief quotations in a review.

T T
551.46
B

LIBRARY OF CONGRESS CATALOG CARD NUMBER: 67–22394

MANUFACTURED IN THE UNITED STATES OF AMERICA

CONTENTS

SOUTH BR.

3

SEAWATCHERS

Oceanographers in Action

1. WATERS OF THE SEA

EVEN though the wind came off the land from a little west of north, it brought no warmth. It was a late winter wind and on that morning in April, there might easily have been a snow squall or two. Lowering gray clouds moved fast over the stack of the ship and the smoke from its idling engines streamed horizontally out across the sound separating Woods Hole, Massachusetts, and Martha's Vineyard, the large island forming the southern boundary of Vineyard Sound. Except for that smoke and the endlessly turning radar antenna, the ship was motionless, lying to at the dock, readying for departure.

A small slate blackboard was propped against the after rail of the mess deck, just beside the gangway. On it were painted the words "Woods Hole Oceanographic Institution" and beneath that identification appeared the information:

Vessel: *R. V. CHAIN*
Cruise No.: 58
Destination: Open Sea
Liberty Ends: 0800
Departure: 0900

A number of scientists stood on the open afterdeck. They wore windbreakers of all varieties—and those with hoods had their heads covered against the cold and the wind. On the dock a few feet away stood wives and children and one or two scientists talking quietly before the time of departure. As the minutes ticked away, cars drove onto the dock and stopped. Men with suitcases, bundles and bags emerged to walk hurriedly toward the gangplank and the ship. One man carried a long fishing rod, for which he was vigorously hooted at by his colleagues. He simply grinned and threatened to bring back a fish.

Forward of the open fantail where the men stood, and beneath the sheltering boat deck, were receptacles and boxes and gear of all kinds. They were the belongings of the various scientists each concerned with his own project. Pails and plastic storage containers, formaldehyde solutions and nets, jugs of distilled water, closed boxes with microscopes, supplies of tubing and pipettes, bottles of test solutions, batteries and spare parts—all the paraphernalia needed for the research projects. Each scientist was acutely aware that if he forgot any item, he could not rush back for it when the hawsers were loosed and the *Chain* had begun to back away from the land, letting the waters of the sea swirl and widen between the ship and the dock where the women and children and other early-rising friends of the scientists aboard became smaller and smaller until they were at last indistinguishable.

Which began to happen exactly at nine o'clock.

The gangplank was hauled aboard and an officer appeared on the starboard wing of the ship's bridge. He glanced first forward and then astern. Waiting crewmen standing by the lines kept their eyes fixed on him. The engines below began to throb more insistently and their power could be felt by those standing on the steel deck plates. The scientists' feet quivered in response to them. The officer on the bridge nodded his head briefly to the men handling the after lines. They flipped the bight of the great hawser from the piling and began to coil it on deck. The stern swung away from the dock in a swirl of wash from the ship. The officer turned and waited, judging the motion of the vessel for a few moments, then he nodded to the men on shore who loosed the forward line and scampered along lifting the hawser over intervening pilings until they were opposite the afterdeck. As the water widened between ship and dock, they heaved the line to waiting hands on deck.

The ship's horn sounded a long blast that silenced everyone. The bow turned toward open water and a great arc of foaming wake curled outward from under the ship's stern. There was another final blast of the horn that echoed across the buildings of the Institution, across the still nearly empty streets of the village itself and up into the low hills surrounding the town to the ancient clapboard houses of whaling captains, houses topped with white-railed widow's walks. It was only another departure. But any departure, whether of short or long duration, in sail in 1800 or under electric-diesel power today, still remains a departure. The wives and friends and children on shore became smaller and smaller until they were only waving memories, while the waters of the sea widened between ship and shore.

* * *

The Research Vessel *Chain* is a converted salvage tug formerly belonging to the U.S. Navy. Her engines can produce 3800 horsepower and she had, in her former existence, towed vessels as large as medium-sized aircraft carriers. Her length is 215 feet with a beam of 48 feet. When the Oceanographic Institution got her, workmen tore out interior bulkheads creating more space for seagoing laboratories. Housed beneath the after end of the upper or boat deck is the great cable drum and auxiliary engine to operate the towing winch, but now the drum holds more than 15,000 meters (over 49,000 feet) of steel cable. This is enough to lower a dredge or other sea exploration equipment to the deepest depths of the blackest abyssal canyon in the sea floor anywhere in the world. A lighter winch and engine are fixed to the starboard side of the boat deck, aft, where light instruments can be let over the side and where nets can be streamed to collect minute forms of ocean life.

The *Chain's* afterdeck, where the scientists had stood awaiting departure, is broad and open. Steel A-frames jut from the stern and the starboard rail. Mounted atop a heavy steel pedestal is a power crane, its long boom lashed horizontally unless in use. Like all tugs, the *Chain* has a low freeboard, and standing on the working afterdeck, the foaming waters of the sea are close to everyone.

As the *Chain* cleared the last harbor buoy, the officer in command set a course down Vineyard Sound. The water was textured by a light wave pattern, but these waters are protected by Martha's Vineyard to the south and the Elizabeth Islands to north and west. Within minutes the wooded hills of Martha's Vineyard's central and eastern areas gave way to the rising red and yellow sand cliffs of Gayhead—the westernmost tip of the island. When the *Chain* came abeam of Gayhead Light, the first long roll of water from the open sea passed beneath her bow.

* * *

On each cruise of an oceanographic ship, one scientist is chosen to co-ordinate all the scientific activity and to confer with the captain of the vessel if any change in plans becomes necessary due to weather, equipment failure or ship malfunction. He is senior scientist for the voyage. On the *Chain* he occupies a large cabin forward on the port side opposite the captain's large cabin adjoining his on the starboard side.

Cruise 58 had been planned to gather data for a number of projects of institution scientists. In addition, several fledgling oceanographers in the form of graduate students were on the cruise, either to take data for a thesis or to gain experience at sea in their chosen field. On this trip the senior scientist had a good deal of co-ordinating to do. Thirty minutes after the *Chain* put to sea, he called a meeting of all scientific personnel in the ship's library where each project leader told what he planned to accomplish. Some had to take their data at night; others, conversely, required daylight for their observations. It thus became obvious that the schedule had to remain flexible throughout the voyage.

While the meeting progressed, the ship's captain put the *Chain* on her course—a course termed by the scientists "the Bermuda transect." This is a line nearly straight across the sea floor from Gayhead Light to the island of Bermuda. In a curious way it illustrates a difference in the way oceanographers think of the vast expanse of the sea and the manner in which most people and all seamen think of it.

The captain, for example, laid his course using the traditional surface divisions of latitude and longitude. He had all the modern navigational equipment and at any time could check his position by Loran—Long Range Navigation. To do this, he would send a radio signal out to established shore stations within range. They will immediately radio back his position based on quick triangulation. Or he could take a sun or star shot with his sextant. Lacking these, he could calculate his position by dead reckoning. In all cases, his attention would be fixed on the surface of the earth—or above it. Not so the oceanographer. He sails the seven seas in a mental glass-bottomed boat. He thinks downward. To him, he is sailing over ranges of mountains, across deserts and plains, canyons and great rivers. When he returns, and is asked where he has been, he may very likely say "on the southwest slope of the Blake Plateau at 500 meters."

To the non-oceanographer, it has little meaning, but the Blake Plateau, lying to the east of Florida, is a well-known feature of the sea floor, and such a description of where one had been would be instantly intelligible to another oceanographer. The statement, "at 500 meters," means that from the surface of the sea to the ocean floor is a vertical distance of 500 meters.

On the bridge, the *Chain* carries a fathometer. It registers the distance from the bottom of the ship to the sea floor at all times, and this information is recorded on an unrolling sheet of paper. The result is a profile of the sea floor along the line of the ship's travel. With his surface navigation and his fathometer, the captain would steer his course until reaching the desired depth. Then with engines stopped, and the ship rising and falling in the swells of the open sea, the first work of the cruise would begin.

But that first station was far off. The meeting of oceanographers broke up and the scientists moved away in twos and threes to discuss their projects. Now land lay far astern over the horizon. No matter in which direction anyone looked, there was nothing to observe but the sky and the clouds and the encircling sea.

THE ORIGIN OF THE SEA

No man can sail for long out of sight of land without being awed by the sea. The smaller the vessel, the more quickly does such a feeling occur. And even the least imaginative of men will eventually ask himself some fundamental questions: When were the seas formed? How were they formed? How was the earth itself formed?

The questions are not new. Men for centuries have been posing them and many ingenious and fanciful theories have been believed. Today, even with much more evidence and many more instruments to measure and work with, men have not answered those questions finally and to everyone's satisfaction. But the decay of radioactive elements, information from stars and planetoids, plus analysis of meteorites that survive our atmosphere as they stream earthward, have all combined to help us frame some reasonable replies to the questions. The age of the earth is presently believed to be between four and six and a half billion years. How the earth was formed —and how the depressions or basins formed that now hold the oceans—is still intriguing scientists of the earth.

One of the currently held theories is that the earth formerly was a mass of hot gaseous material that had been ejected by an explosion on a star such as our sun. It may well have been our sun.

Gravitational and magnetic forces attracting various particles of the hot gaseous elements and compounds gradually pulled them closer and closer together, until at last a sphere of violently hot and bubbling liquid became the planet earth. As the intense heat of the earth radiated away into space, a number of things occurred. Heavier materials "sank" to the center of the earth. Lighter material in the liquid ball floated on the heavier material just as oil floats on water. This lighter material eventually cooled enough to harden and become a solid crust. It still was shot through with erupting volcanoes, and the heat from the interior broke the crust again and again, thrusting material out in unbelievably large eruptions. But now, with a crust formed, the surface cooled more rapidly since the crust lost heat to outer space through radiation faster than it came from the interior by conduction.

As centuries passed, the cooling crust shrank. Like a withered apple or orange, the "skin" wrinkled. Some of it wrinkled outward and some inward. Where the skin was pushed out by shrinkage, we have the continents of the earth, where it shrank inward, the ocean basins.

This is, at the moment, a currently accepted theory on how the earth was formed.

How the oceans themselves formed is less a matter of guesswork.

All the known naturally occurring elements on the earth were present in its formation—from hydrogen, the lightest, to uranium, the heaviest. We know this because, otherwise, they could not be here today. Some of these elements—in fact, the lighter, least chemically active of them—were probably here in greater quantity at the time of the earth's formation than they are now. Helium, for example, was first discovered and fitted into our list of elements by observation of the sun. Only after it was identified there by its particular light emission was helium discovered in small quantities trapped in rock deposits beneath the surface of the earth.

Helium is a very light gas and is chemically among the so-called inert elements which include argon, neon, krypton and xenon. Hydrogen, as the lightest gas, should have escaped from the cooling and solidifying earth except for the fact that it is anything but inactive chemically. It combined with oxygen to form the familiar water molecule. Thus, at the earth's forming, there was a great deal of hydrogen converted and, therefore, trapped here in the form of water vapor.

This vapor hung over the surface of the hot earth for millions of years until the crust cooled enough for it to form clouds and condense into liquid water. Then the entire earth was subjected to a deluge of rain that surpasses our understanding. For hundreds and hundreds of years rain fell to the earth, only to be turned immediately back into steam, just as water droplets falling on a hot stove immediately boil away. But as the steam rose to great heights, it cooled again and fell once more as rain, again and again, until at last the surface of the earth had cooled enough to allow the rain to stay as water on its surface. Responding to the pull of gravity, the water poured from the upthrust continents, flowing downhill to the ocean basins. As it swept across the jagged granite faces of the continents, it dissolved some of the rock, taking elements and compounds with it that would remain forever as the salt of the young brackish sea.

Yet this, according to some oceanographers, is not the

manner in which the sea obtained the majority of its salt. Tests in the fresh water flowing seaward near the mouths of rivers indicate that most of the dissolved compounds there are carbonates, while the greatest percentage of a "salt" in the ocean is a chloride. Using arguments like this, oceanographers can maintain that the original salting of the sea came from the dissolving of compounds in the crust of the ocean basins when the seas first formed. Yet it is true that rains still wash away the land and carry to the sea certain dissolved salts that contribute to the sea's salinity.

It is believed, for that matter, that all known elements could be found in sea water. Many, however, are present in such small quantities that their presence is difficult to detect by usual chemical analysis. And here the myriad forms of sea life come to the assistance of an oceanographer bent on discovering the elements in the sea. Some sea creatures extract these minute traces of elements from the water and use them in their body structures. Analysis of these animals affords the scientist a glimpse of a tiny concentration of the element sought. Lobsters, for example, extract copper and cobalt, ascidians have vanadium in their blood, and nickel has been detected in mollusks.

THE EXTENT OF THE SEA

When an artist wants to give the visual impression of infinity—of endlessness—on his restricted canvas, he frequently uses long horizontal lines, and it is this sense of endlessness that impresses anyone standing on the deck of a ship, gazing at the watery world around him. The horizon is essentially a straight line to such an observer, and if he is some 30 feet above the surface, he can see less than 150 square miles of restless water.

Nonetheless, the immensity of the sea presses on the imagination. It is not a false impression, for the sea truly is immense. It covers more than 70 per cent of the earth's surface and there are 350 million cubic miles of it to consider. In high southern latitudes, a ship could sail forever circling the earth and never sight land. The average depth of the ocean, including many of the shallow seas, is more than two miles. Its greatest known depth is more than 11,000 meters. This great depth is located in the Mariana Trench in the Pacific Ocean. There is so much water in the oceans that if all the continents were washed into the basins to make the earth's surface completely smooth, the oceans would cover the globe to a depth of nearly 1.4 miles.

These are the obvious facts about the sea that produce awe in any observer. There are others less easily measured or understood. Sea water, for example, has a higher heat capacity than any common material except liquid ammonia, that is, it can absorb more heat energy with less temperature rise than other liquids. It also has a high electrical conductivity, which produces one unfortunate problem for men: radio waves will not travel far or clearly under the sea. The result is that researchers have concentrated their efforts on sound waves as a means of underwater communication. Today acoustical apparatus is used to send signals outward for communication or to have them reflect from objects to become underwater radar.

In living by and on the sea, man has taken its existence for granted. But our recent observations—some at close range—of other planets force us to admit that within our solar system, at least, water is a strange substance. If beings on Mars, or some other planet, were to observe earth closely, they would have to conclude that Earth is a very wet place. Yet water is so common and familiar that we scarcely realize its uniqueness in the observable universe.

Consider these facts about water: It exists as a liquid within a narrow temperature range, from 32° F. to 212° F.—and if the temperature of most of the surface of the earth did not fall within that temperature range, no life as we know it could exist. When it evaporates, it rises into the atmosphere to condense and fall as rain, making lands fertile to produce our food. Unlike other liquids, water has the peculiar characteristic of expanding when it freezes. Ice is lighter than water and, therefore, floats. If it did not float, much of the world's population would be drowned by seas rising above unmelting ice that would sink to the ocean floor.

MAN'S VIEW OF THE SEA

Since before recorded history, the sea has influenced man's life on earth, both physically and emotionally. His curiosity and sense of adventure sent him out in small boats far from shore. Many accounts of ancient voyages indicate that men migrated across intervening oceans to settle new lands. Men have also learned to draw much of their food supply from the sea. While prehistoric men did not realize it, the sea acted to make lands adjacent to it habitable by modifying the climate. The greatest cities of yesterday and today were built on the shores of the sea, which became the great highway for trade, expansion and wealth. Yet only recently has man made any concerted effort to analyze more than the surface of the sea, or the waters adjacent to his terrestrial home.

From the poems by Homer to the pages of Melville's *Moby Dick,* the romance of the sea as well as the fear, reverence and awe it inspired have been part of man's literary and aesthetic heritage. Mermaids, sea monsters, Jonahs, legends, poems, tales of great voyages and storms have all filled man's imagination.

Such fancies should not be surprising since all life on earth began in the sea. Its salt is in our blood and men who live near the sea find that its tides and moods control their lives in practical and psychological ways. The spread of mankind to the four corners of the earth has been made possible by the great highway of the sea, the original unifying link of all mankind.

No writer of ancient times or the present has put the matter more convincingly than Herman Melville. In one chapter of *Moby Dick* he writes half in fun and perhaps more than half in earnest of the odd manner in which mankind searches for water—not to drink, but to be near, to see and to feel. He writes of the crowds that throng the beaches, of the hypnotic effect of the waves drumming on the shore. He mentions with pity those who are far from the sea and can only huddle around lakes and streams. But huddle they do, and they will sit for hours musing at the shimmering surface of the water, held unconsciously by some small but vital shred of man's former beginnings with all life beneath the constantly moving surface of the sea.

And man has named the "seven seas," so to speak. To the ancient Phoenicians the Mediterranean was a well-known sea. These hardy mariners sailed through the Pillars of Hercules and explored the eastern shores of what we now call the Atlantic Ocean. Historians believe these early mariners sailed as far north as England and westward out of sight of land to the Canary Islands and beyond to the Sargasso Sea. The Greeks themselves were sailors, though of a less daring sort. Yet their mariners brought back stories of faraway lands in the Atlantic. Within the history of man, the shapes and locations of the "seven seas" have remained much as we know them today. With relatively minor changes in shoreline, these bodies of water (actually linked, and all a world ocean) have been charted, but we know they were not always in the same place or shape.

Long before man evolved to his present form, the land masses and sea basins occupied quite different parts of the earth's surface. The seas have shifted because the crust of the earth itself has not remained as it first formed. Through millions of years it has buckled, pushing up mountain ranges and pouring water into the resulting depressions. In a way this movement of the earth's crust answers the question: Why has not the land of the continents washed into the sea as a result of the rains that have fallen throughout ages of time? If the earth's crust had remained motionless, this probably would have occurred, and the entire face of the planet would be covered with water.

But the earth's crust *has* moved. This movement, and the tracing of ancient sea basins, belong to the study of palaeogeographers. One of the ways they trace the ancient sea floors is by examining rocks on what is now dry land and finding fossils of forms of life that only live in the sea. Such a discovery means that the land there, at one time or another, lay beneath the sea. For one example, we can examine their findings of the Cambrian era—approximately 500 million years ago. Fossils of sea animals known to be alive at that time have been found in many unlikely places. The oddest of these is on top of the greatest mountain range existing on earth today—the Himalayas.

During the Cambrian era the area between South America and Africa was dry land. This stretch of land reached east all the way across what is now the Indian Ocean and included the western coast of Australia. This vast land has been named by palaeogeographers "Gondwanaland." The center of the Pacific Ocean was still ocean, as was the North Atlantic, but all of what is now western Europe lay under water, and water stretched from northern Mexico northward across the western plains of the United States into the Arctic basin. India, the Himalayas and China were drowned lands, lying beneath what is called the Redlichia Sea. Central Russia lay under water while much of what is now the East Indies and the southwestern Pacific Ocean was all dry land.

Except when earthquakes occur, such wrinkling of the earth's crust and changing of the ocean basins require millions of years. Even the mightiest earthquake in recorded history is a tiny ripple compared to these long-term earth crustal movements. Yet in the world as you know it today, there is evidence of crustal movements occurring. The coasts of Maine and Norway are called "recently drowned coasts." The land subsided, perhaps pushed down by the unbelievable weight of the ice that was present during the last ice age. As the ice melted, water from the sea flooded the land that formerly had been above sea level. Thus, the wooded, rocky fingers of land reaching out into the sea are all that remain of high hills—just the crests sticking out of the water.

Great rivers, carrying mud and bits of rock to the sea, slow down as the fresh water meets the oceans. As the rivers' currents slow, mud falls to the bottom, building great delta systems. Over the centuries the increasing weight of the delta material pushes the earth's crust down, and once more the oceans flood the land. An outstanding example of this kind of drowned coast is Chesapeake Bay.

The sea encroaches on the land in other cases and for quite different reasons. Most such examples of the rising sea are as slow or slower than the examples given above. Water, trapped in the interior of the earth since its formation, is released in volcanic explosions, either beneath the sea or in the earth's atmosphere, and it adds to the existing supply. There is another source of water to make the seas rise: polar and glacial ice on the Antarctic Continent and Greenland. There are, of course, other places where glaciers exist and where more or less permanent land ice remains intact. But if the ice from the Antarctic Continent (which in places is over two miles thick) and that of the Greenland icecap were to melt, the seas would rise over a hundred feet above their present levels. This means that all the cities and towns on the eastern coast of the United States would lie under water. The seas would reach over all that coastal plain, and the shore would be somewhere in the foothills of the Appalachian Mountains. The low-lying Gulf coastal plain would be flooded and cities on the West Coast inundated.

The sea actually is rising, although not as a result of melting ice to any great extent. Over the past 35 years, the seas surrounding the North American Continent have risen several inches. The cause is hidden from us by time—the thousands of years that must pass before the pattern of change reveals itself.

ENERGY IN THE SEA

No one who views the sea either from the shore or the deck of a vessel like the *Chain* can hold the idea that the sea is a static body of water—motionless and passive. The sea is forever in motion, and some of its motions have only recently been discovered. The motion of waves and tides and currents gives to the sea incalculable mechanical energy, the kind capable of exerting a force upon any resisting object, whether a boat or a continent. Unfortunately much of this energy cannot, from man's point of view, be turned to useful work.

Some of the energy has been calculated, however, and men have dreamed, and still do, of eventually putting more of it to work operating generators to supply electric power for their needs. In Europe today there are several generating plants that harness tidal forces to produce electric power. And the decades-old plan to use the power in the extreme tides of Passamaquoddy in the Bay of Fundy is still in government files. As men learn more about the technical problems in such a large project, Canada and the United States may benefit from its completion.

The energy of waves has fascinated man for centuries, principally because he has watched as giant breakers crash against the seacoast and dissipate their energy on the land. Such waves can strike the shore with the fantastic pressure of two tons per square foot, but so far men have not been able to make use of these great forces.

There are also currents in the sea—great ocean rivers, some of which are in the upper levels of the sea and, like the Gulf Stream, can be distinctly noticed by man as he sails across its surface. Others lie at medium depths, and still others make their secret ways across the bottom of the oceans all over the world. The hope of using the energy of any of these directly, to do more than assist in moving our vessels through the water, is remote.

The other form of energy in the sea is heat. Unlike the pattern of waves and tides, this energy is not readily visible and so it has escaped man's notice for a long time. The great world ocean is actually a reservoir of heat of such vast proportions that we have not measured it, much less been able to put that energy to direct use.

The two sources of heat are the sun and the decay of radioactive materials in the earth's crust beneath the seas. These radioactive sources of heat have only recently been discovered and some of their effects observed. But certain areas of the sea floor have temperatures higher than others, and this difference causes warmer, less dense water to rise, while colder water flows in from the sides and bottom of the resulting column of warmer water. Thus, slow-mixing currents of water take oxygen down to the deepest parts of the ocean and help support the life that exists there.

The sun, however, is the principal source of heat for the oceans. The sun's energy either finds its way directly into the surface layers by absorption, or it heats the surface as sun-warmed winds move across its face.

So far man has been unable to harness the heat energy in the water, but it is there and in such great quantities that if one cubic mile of sea water had its temperature lowered just one degree Fahrenheit, enough heat would be released to turn 132,000,000 pounds of water into steam. There are 350,000,000 cubic miles of sea water in the oceans of the world, and they constantly receive heat from the sun, so the heat supply is endless. Despite the fact that we cannot make direct use of this tremendous storehouse of heat energy, we receive considerable benefit from it through the action of ocean currents that move from warm tropical waters to dry or frigid land masses.

The flow of warm water into the North Atlantic Ocean (coming principally from the Gulf Stream) so tempers the climate of England and the coast of Norway that those lands have a more temperate climate than they otherwise would. This means that men can live comfortably in those lands, and the temperatures are moderate enough to allow a growing season for plants. Thus there is fertility enough to support man and the forms of life he depends on for food.

Some portions of the earth's land masses, notably India, receive another benefit from the heat absorbed by the sea. Warm waters of the Indian Ocean evaporate, and the water vapor saturates the air passing over it. During the monsoon season, prevailing winds move this water-saturated air far inland where the moisture condenses and falls as rain. Without this water, both for drinking purposes and for the growth of crops, parts of India that today support large numbers of people would be deserts.

So, even without man's effort, the sea makes life possible for millions of human beings who have never gazed on the great oceans.

2. THE EARTH BENEATH THE SEA

As the *Chain* moved farther from land, she became more exposed to the great mass of cold northern air that had chilled the scientists that morning. The temperature dropped to 38° and the wind freshened. From the bridge, the captain sent word down that the ship would be on the first station at midnight.

The scientists had much to do in the laboratories in preparation for the work ahead. The largest lab was just abaft midships on the mess deck. On the port side of this protected space, a long bench had been constructed and electrical outlets placed for any equipment an oceanographer might bring aboard. In one enclosed space, the biologists' microscopes, lights and camera stands were set up. On the boat deck above, another laboratory housed standard water-sampling equipment. Here one scientist was busy arranging his sampling bottles, cleaning glass tubing and checking valves to be certain they did not leak. He placed his test solutions carefully in a wooden rack so that no pitching or rolling of the ship would send them crashing to the deck. In yet another high laboratory over the bridge, a PGR, or Precision Graph Recorder, was turned on. It functioned as a continuous water-depth indicator or fathometer. The scaled graph paper began its slow unrolling and the "pinging" of the device broke the silence as the long wavering line began to chart the bottom of the sea over which the ship passed. A water surface temperature reading machine gave a continuous account of the temperature to hundredths of a degree, by means of infrared radiation.

The course steered by the ship was not new to the men on board. In fact, the same group of scientists had made the "Bermuda run" many times. They had good reason for this apparently needless repetition. By sampling sea life and taking measurements over the same path, they felt that eventually they could put data together and have a reliable profile or cross section of a typical portion of the sea. From the picture they would draw with the information obtained, they could, with some qualification, generalize about certain characteristics of the sea anywhere in the world. Some of the questions they were trying to answer included: How much life existed on the floor of the sea? Was the salinity of the water constant throughout the year? Throughout several years? Did the temperature at various depths change much? Or at all? Was there a measurable difference between summer and winter conditions in any or all forms of life found in the seas there? The answers to these and other more complex questions would go to build the picture the scientists wanted. So this voyage was simply one that fitted into a large pattern of research.

In addition to putting the laboratories in order, each scientist had to get his personal gear below to his assigned cabin. Most of the lower cabins had four bunks. Two smaller cabins above on the boat deck held two bunks each. There the women biologists were quartered.

Darkness fell that day without a break in the heavy overcast. A quick turn around the deck showed the sea rising and the wind now strong enough to send the crests of waves flying into the gathering dark as white bits of foam.

At dinner the captain confirmed the rumor that far ahead, beyond the first station, a storm was whirling over the wide sea. Weather reports from shore stations came into the radio room of the ship, pinpointing the disturbance. The only concern the scientists felt was for their work. If they could reach the first station before the sea rose too high, they had a chance to get their data. In a way it became a race between the *Chain* and the approaching storm.

At this time the *Chain* was sailing over one of the three major divisions of the sea floor—the continental shelf. Although this important section of the earth's crust was not of primary interest to the scientists on this voyage, it is a region undergoing intensive study by other groups, and it is probably the most important area of the sea so far as man's immediate needs are concerned. The word "shelf" does not convey the actual shape or appearance of this part of the ocean bottom, although the term is accurate. Save for the water that covers it, the continental shelf resembles another feature of the earth's crust —one that all Americans have some knowledge of.

What eye, young or old, has not seen—at least on a movie or TV screen—the endless rolling plains of the western United States? There, limitless and undulating to the horizon (perhaps with covered wagons struggling across it) lies nearly flat grassland. Now imagine these plains existing beneath the waters of the sea and near the edge of the land. They are the same in appearance, endless and nearly flat. But these lands are called the continental shelf. They extend, on an average, 30 miles from the shore's edge. In some places they do not exist at all and the land plunges to great depths just offshore.

In the Arctic Ocean, however, along the coast of Siberia, these flat continental shelves reach 800 miles out from the land. Oceanographers generally designate the edge of the continental shelf at a depth of about 200 meters. If you were on the shelf, at the bottom in an undersea craft, you would not be able to tell whether the bottom sloped downward or not, for the eye cannot measure the gradual falling away of the shelf. The drop is around two meters vertically for every mile traveled horizontally.

It is over this great drowned coastal plain that most of the life of the ocean exists—over it and on the bottom. For into depths up to 200 meters some sunlight can penetrate. It is not all "pure" sunlight, to be sure, for as the white light of the sun penetrates the waters of the sea, some of its wave lengths are absorbed before reaching any appreciable depth. Red is the first color absorbed by the sea. Then the other visible wave lengths of light are absorbed until at last only blue remains at the opposite end of the light spectrum from red.

Yet, employing whatever light comes to them, the sea plants, like plants on land, use light energy to produce growth by photosynthesis. In this process, carbon dioxide found in water and the water itself are chemically combined by the light energy to produce a basic organic material for plant growth. A by-product of this complex reaction is oxygen, which the plants discharge into the sea. Not at all by accident, animal life, both terrestrial and oceanic, needs oxygen to survive and it gives off, as a by-product of respiration, carbon dioxide. So the cycle continues.

On the continental shelf and especially in its shallower waters, sea plants of all kinds abound. These vary from the tiny one-celled diatoms which, like minute animal forms, "grow" by cell division, to the great waving strands of kelp attached to rocks and the soil of the bottom in shallow water that swing back and forth in response to the waves and tides of the ocean. Small unattached plants, like the diatoms and dinoflagellates, drift everywhere in the sea. They must, however, stay near enough to the surface to receive their required amount of sunlight to sustain life. Such plants that drift with the currents are called phytoplankton, from the combination of two Greek word forms. The prefix "phyto" means plant, while the word "plankton" comes from "planktos," or wanderer.

These small plants are the primary organic key to all life in the ocean. The smallest animals—zooplankton or animal wanderers—feed on the lowly diatoms and dinoflagellates. The zooplankton in turn are eaten by larger fish, and so the entire cycle of oceanic life regulates itself.

Since most life and the easiest fish to catch are found on the continental shelf, men have for untold centuries put out in boats and with nets, seines and trawls, or a series of many-hooked lines, managed to gather a harvest from the sea. In olden times (even now in small primitive communities) such fishing did not upset the cycles or the number of any specific fish species. But as man's technical skill grew and his own population exploded, the problem of finding enough food stimulated his ingenuity and produced the great modern fishing fleet.

The great fisheries of the world have traditionally been in the North Atlantic, off the Japanese coast, off Malaya and the South African coast. Newer fisheries have sprung up off Australia and the southeast coast of South America. In Japan alone, for example, two million people make their living by catching fish or tending to some aspect of the harvest of food from the sea.

Such intense fishing as we now do in many areas of the ocean has resulted in a marked reduction in some species of fish sought by man. Herring is a good example. This fish has been found and caught the world over. Yet today, in many areas where herring fleets are numerous, the annual catch is falling. Off the coast of Maine, for example, herring fishermen use an airplane to try to spot the schools of herring that formerly were so abundant in those waters.

At first glance it would not seem that man could possibly overfish the oceans. In some species the female will lay as many as three million eggs in spawning—yet before the fish hatch and grow to maturity, all but two of those eggs or young fish will, on the average, have been eaten by other sea life. So, with man's interference, the supply dwindles and the former big catches and large specimens of the species dwindle until it becomes no longer profitable to fit out boats and hire crews to put to sea.

A fisheries biologist, searching for ways to conserve the fish that can feed mankind, studies the species under consideration. He may discover that the eggs of the female must be left undisturbed in waters of medium depth, and so he tries to prohibit trawling nets from being dragged along the bottom during the spawning season. Or he may study a species of fish that requires several years to mature enough to be worth catching, so he will argue for nets with large enough openings between the strands to let the younger fish swim through and escape to grow larger.

But wherever he turns, the biologist will run into difficulties. Shortsighted men wanting a profit will fight all attempts to regulate their fishing. Until at last the species is so reduced that even the greediest fisherman is willing to call it quits and submit to regulations.

Another influence of man now threatens many forms of plant and animal life on the continental shelf. This stems from the pollution that is entirely man-made. Between fifty and sixty million people live within 50 miles of the coastlines of the United States. The discharge of sewage into the coastal waters has already depleted the marine life along the coasts where the discharge is greatest. Today there are government biologists working to analyze the problem. Oyster beds in many places that once offered a bountiful harvest to men are gone. In the estuaries of large rivers where industrial waste and untreated human sewerage is highly concentrated, fish have become a rarity. The estuarial situation in some areas has prevented certain species such as salmon from returning to their spawning grounds.

Fallout from radioactive waste is another hard problem confronting scientists. Strontium 90, a particularly slow-decaying radioactive substance, has been found in the bones of fish. How much of it is present in the flesh of fish we eat is not yet known. Plutonium 239, a man-made radioactive material that requires over 24,000 years to lose half its radioactivity, has been found secreted in the shells of lobsters.

The pollution and the overfishing have occurred principally in the region of the seas called the continental shelves. Physically, the true edge of the shelf is exactly that—an edge. For the gentle, imperceptible slope does end and a much steeper slope leads rapidly down to the sea floor. This slope is much like that of a rugged mountain range, or a cliff. From 200 meters it ranges downward to 3000 meters, and its flanks are scarred and ribbed. The scars and the intervening ribs are not small. Some are huge canyons as large as the Grand Canyon, and the intervening ribs jut out to form great bluffs.

The continental slope undergoes frequent and awesome changes. It is also the source of some disturbances that affect men thousands of miles away. On occasion, in the deep silence of the ocean, fragments of the "cliff" break off and plunge to the sea bottom. If we could observe these occurrences, we would probably have the feeling that we were looking at a silent motion picture being run in slow motion. The water tends to support the falling fragments, and, when huge sections split away, they must plunge with a slow, dreamlike quality, turning end over end and vanishing into the darkness below. A less dramatic though equally important change occurs when the slope is covered with sediment either from currents depositing river mud or from the steady slow rain of dead organisms building up layers over the centuries. When enough mud or sediment has accumulated on this precipitous slope, a gigantic land or mud slide can occur. It begins with the first fall of a few particles, tipping the final unstable balance, and again, in eerie silence, the mass slides along the slope to the dark abyss.

When such major changes take place, tremendous pressure waves are created that travel through the water, for the slides or breaks on the continental slope are explosive releases of energy. Much of this energy is carried by these invisible waves of pressure beneath the surface. The waves are invisible to everything except a pressure-measuring device. If such a device were in place at the time of the undersea cataclysm, it would fluctuate wildly as walls of water pushed outward on each other. Like a giant expanding bubble, the first pressure wave would break to the surface, piling water upon water and creating what most people would call a tidal wave. The Japanese term for such waves is tsunami. Scientists call them seismic sea waves—those caused by earth movements. The speed of these waves is tremendous, reaching as high as 500 miles per hour, depending on the depth of the water in which they occur. The waves thus generated on the surface will travel across the widest ocean in the world and can cause great damage and loss of life on distant shores. Curiously, these seismic sea waves are not often noticeable at sea. Passengers on a ship may travel over one without being aware of its passing, for the distance between the relatively undisturbed surface of the water and the crest of such a wave may be several miles. Its forward slope is so gentle that the eye cannot detect it.

Frequently earthquakes, well below the earth's surface beneath the sea, trigger the enormous break or slide on the continental slope. Sometimes it is an event so small in comparison that it can never be detected. Yet, when it occurs, it is the final straw that breaks the back of the slope. Gradually, the final bits of earth and rock subside, the cloudy waters clear and the continental slope in that particular area has a new shape—for a time at least.

The third major division of the earth beneath the sea

is the sea basin itself. Depths here are between 3000 and 6000 meters. These are nearly constant the world over, but by no means are these basins to be thought of as featureless plains. Huge mountain ranges, flat-topped mountains called guyots (gee-ohz), deep trenches or canyons going to the deepest point beneath the water surface, like the Mariana Trench, reach down into true abyssal depths as far as 11,000 meters.

It is on the floor of the ocean basins that material falling from the continental slopes finally comes to rest. Other material finds its way to the sea floor, too. The ocean basins are covered to an average depth of 300 meters with sediments of various kinds. Much of it is organic, being the dead bodies of countless sea creatures that have been falling to the sea floor since the beginnings of life itself. Another type of sediment comes from small quantities of silt and clay from the surface of the earth above water, washed down by rivers or carried by wind. Most of this type of sediment, however, is found on the continental shelves or slopes. A third type of sediment comes from glaciers. When these vast sheets of ice retreated, they left behind material scraped up from other places and dragged along with the ice sheet itself.

The fourth kind of sediment is volcanic in origin. Although some of this came from volcanoes on land, the ashes carried thousands of miles by the wind, much of it found its way to the ocean floor as a result of undersea volcanic explosions, for these eruptions from comparatively weak places in the earth's crust occur beneath the sea as they do on land. Lava and other products of volcanic action have been found in the depths of the sea.

If we now had an undersea craft that could start from the shore and cruise right along the sea bottom to the abyss, we would find first the wide, gently tilted continental shelf. At first there would be plant life clinging to the bottom, but as we went deeper, there would be less light and fewer plants. Thousands of fish in schools would flash by our craft. If we glanced down at the bottom of the sea, there we would see the typical bottom fish, the flounder, the skate, the lobster and the crab, with here and there an eel.

As the darkness deepened, we would approach the edge of the shelf much like the edge of a cliff. Here also would be fish by the thousands, for many schools patrol this region of change. Plankton, too small for us to see from our moving craft, would be the prey on which these fish were feeding. If the conditions were right, we might see many small shrimp propelling themselves along with vigorous motions. At the edge of the shelf, if we turned on our lights, we could look over at the rugged slope reaching down into the depths, and by pointing the nose of our craft down along it, we would descend, dodging outcrops of rock and sinking farther and farther until we reached a depth of 3000 meters or more.

At the base of the continental slope we would find a long ridge of material running parallel to the foot of the slope—the accumulation of sediment that has over the centuries fallen from the slope itself. Then we would be on the floor of the sea basin. Perhaps we would first see it, in the rays of our searchlight, as another limitless plain, for some parts of the basin are as flat and monotonous as gymnasium floors. No plants would greet our eyes. There would be no swarming schools of fish, but if we were to rest quietly on the bottom for a few moments, turning the searchlight slowly around us, we might see several bottom fish of odd shapes and sizes. We might also see the simple wriggling shrimp go past us, for this small crustacean has been spotted by undersea craft in the deepest parts of the sea.

If we then proceed along the ocean basin, it will be only a matter of time until we come upon some striking feature of the sea floor. It might be an entire range of undersea mountains rising in the gloom, or a canyon winding its way across the basin. If we penetrate such a canyon or trench, we are reaching into the lowest depths of the sea. We might drop down five or six thousand more meters into this trench that comes closer to the interior of the earth than any place on the planet. On such a journey as this, the *Trieste,* the famous undersea craft, carrying two men, was lowered to the floor of the Marianas Trench. In that eternal darkness at the bottom of our world, the *Trieste* actually located life, a large bottom fish blinded by the glare, and the ever-present shrimp still wriggling its lonely way to Nowheresville.

Rising to the ocean basin once again, our undersea ship could take the fascinating voyage around and over and among the mountains of the sea floor. In the mid-Atlantic, for example, there is the mid-Atlantic Ridge, an entire range of undersea mounts reaching from Iceland at its northern end and running as far south as the tip of South America. Occasionally as we passed over one of the crests, we would discover it to be as flat as a table. In our own West, flat-topped mountains are called mesas, which is Spanish for "table." Under the sea's surface, they are called guyots after the scientist who discovered them. When our undersea journey was completed, we would know without doubt that a world of landscape is hidden from our eyes. It is as varied as the earth above the sea, if not more so. Life exists everywhere in the sea, even though that life is limited at great depths—because of the tremendous pressure of the water and the complete absence of sunlight—to relatively few animals.

Only in the past few years have men begun to chart the infinitely varied landscape of the sea floor. Many surprising features have come to light. In areas thought to be flat and featureless, mountains have been discovered. The mid-Atlantic Ridge we now know is part of a world-circling belt, or range, of mountains interrupted here and there by trenches running across the mountain range and dropping to tremendous depths.

Oceanographers' methods of exploring the nearly 400,000,000 square miles of oceanic floor are increasing in efficiency as time passes, but the sea is not a respecter of men's budgets, and exploration still is a very expensive process. There are many types of small undersea craft now being developed, some for use in shallow water, some in medium depths, and still others like the *Trieste,* designed to be lowered into the deepest trenches in the world. None, unfortunately, has yet been developed that could take the trip we just imagined from the shallow coastal waters to the abyss. Even if we had such an undersea craft at our command now, we would have to build a great many of them and employ thousands of people as undersea explorers to make a serious beginning at covering the vast area of the earth that is covered by the sea.

Unquestionably, the exploration of the sea floor will be carried out more and more frequently by the small maneuverable submersible. Many industrial companies and governmental branches are busily developing these craft. Most of them, carrying two or three men, are self-propelled with small electric motors. They carry a power supply that enables them to use searchlights and other equipment outside the underwater vehicle. Some are "developing arms," and these mechanically jointed and operated extensions allow the explorers to reach out and pick up objects that attract their attention. The items retrieved can, in some cases, be placed in a receptacle in the craft and carried to the surface. In other cases, items wanted for study on the surface are placed in a basket which, when full, is hauled to the surface by waiting assistants.

Such exploration is, at the moment, very expensive. In fact, it will always be expensive since these undersea vehicles must withstand great pressure; must carry supplies, power, oxygen and a growing number of electronic devices to allow them to function in the strange environment of the sea. In addition, nearly all of them require that a ship be standing by on the surface to raise or lower the vessel by means of cables.

Many oceanographic expeditions must be planned very economically. As a result their methods of gathering samples from the sea floor may appear to be both crude and primitive. The methods used have only one thing to recommend them: they work. Clamshell buckets, built on the principle as a kind of power shovel on land, are lowered to the sea floor. When the bucket strikes bottom, its two open jaws bite into the floor. When force is exerted upward on the cable, the jaws are driven shut, trapping the sample so it can be raised to the surface. Another and more common method of getting bottom samples is by towing a dredge at the end of a length of cable dropped over the stern or side of the ship.

This is the method used by the scientists on the *Chain*. The dredge rested on the open afterdeck when the ship left Woods Hole. During the afternoon and early evening of the first day, the men worked putting it in shape and checking it over.

The dredge was a flat oblong box. Its steel frame was covered with heavy galvanized wire mesh and fitted with skids of metal wrapped around the long edges of the box. One end of the box was open and it was framed with two heavy steel lips or flanges set open like the jaws of an alligator. These were adjustable, to vary the "bite" of the flanges on the sea floor. Inside the dredge, hidden beneath the galvanized screen, was a cloth net of nylon, having small regular openings which would catch and hold whatever specimens were trapped by the lips of the dredge. Water could pass through the entire envelope, but the small openings kept even extremely small specimens of animal life or sediment trapped for inspection when the dredge came to the surface.

This particular dredge devised by the scientists had been tested many times, but it still demanded their attention. On many occasions they had waited patiently through the long hours while the ship towed the dredge and then hauled it up, only to discover that it contained few or no specimens from the unseen floor of the sea. Part of this exasperating difficulty they blamed on the fact that as the dredge was being hauled upward, with its open mouth pointed toward the surface, the ship itself rising and falling in the waves produced a "winnowing effect," because the dredge at the end of the steel cable had to rise and fall in concert with the ship. Since waves on the open sea were often ten and more feet from trough to crest, this meant the dredge was being lifted and dropped, lifted and dropped every time the ship rose and fell. As a result, specimens floated out of the open end of the net.

To overcome this problem, the scientists had a nearby equipment company build a spring-operated metal door for the mouth of the dredge. Activated by a timer, it was supposed to close on signal when the tow was completed, and the springs were to hold the door shut to minimize whatever surging took place as the dredge was hauled to the surface.

During the last hours of the evening before the ship reached its first station, the scientists and the representatives of the equipment company fretted and tinkered with the additional mechanism. It seemed to be in perfect working order.

At twelve midnight the *Chain* arrived on the first station at 39° 40′ North Latitude, 70° 35′ West Longitude at a depth of 500 meters. It was a position on the continental slope and one from which the scientists had frequently taken specimens. The senior scientist signaled the bridge that he was ready to lower away the dredge. The winchman was in place at his controls. The men dragged the dredge to the after fantail and snapped on the steel towing cable. The winchman reeled in a few feet of cable which raised the front end of the dredge from the deck. The men hand-wrestled its lower edge over the stern. Light from high-powered floods illuminated the figures and cast long, leaping shadows across the deck. With a wave of his arm, the crewman directing the task of getting the dredge safely into the water signaled the winchman to lower away. The cable drum rumbled and the cable slowly began to pay out. The dredge disappeared beneath the black surface of the choppy sea.

The *Chain* had stopped all engines. She wallowed in the waves that hissed against the hull. The only other sound came from the cable drum across the deck, while the men stood silently on the stern watching nothing but the thin steel cable being fed endlessly into the sea.

The questions in their minds centered on the trap door that had been rigged, on the luck they would have this time with their effort, on the terrain they would strike with the dredge when it hit bottom. There was a great deal to think about but, with the black sea between them and their goal, not one thing could be done to answer their questions.

The cable was paid out at the rate of 50 meters a minute. When ten minutes had passed and the indicator on the winch showed that 500 meters had been lowered, the scientists signaled for 200 more meters of cable to give the slack necessary to have the leading edge of the dredge dragged flat on the bottom far behind the ship.

A signal to the bridge brought the engines to life. The *Chain* turned and began moving slowly at about two knots, running a course parallel to the continental slope, keeping as close to a 500-meter reading on the fathometer as possible. They were dragging along the contour on the side of the undersea cliff. For an hour the dredging continued and the scientists stayed on deck. They drank cups of hot coffee and turned up their jacket collars. The wind was cold. It was also rising. Although they had beaten the storm to the station, they had not won the race by much.

At the end of the dredging time, the senior scientist nodded to one of his colleagues who went forward and stuck his head into the crew's mess, where the winchman was keeping warm and happily sipping coffee. At the sight of the scientist, whose face was red with cold and even a trifle blue, the winchman sighed and tossed down his hand in the gin rummy game. When he had tugged on his foul-weather gear, he took his place at the winch controls. The scientist signaled the bridge, and the ship's engines stopped. The cable drum growled and the cable began to come up.

By the time the dredge broke the surface, all the project scientists were leaning eagerly over the stern, hoping to be first to sight evidence of a good haul.

"The door's shut," one called.

The equipment man from the company sighed in relief.

"We got something," another shouted.

Then eager hands reached out to guide the heavy steel frame up onto the deck. They had indeed "got something." In the load of bottom sediment were the red legs and claws of crabs, a fish tail and fin wriggling in the unaccustomed lump of sand, and various squirmings from the surface of the sand itself.

When the winch deposited the dredge on the deck, the bottom of the dredge net was opened. One man brought over a large galvanized container and one of the scientists began hauling sand, rocks and specimens out, being careful to have everything fall into the container. It was cold work. Not only was the wind whipping across the deck, the waves covering the men with spray, but the sand into which the scientist again and again plunged his hand was only a degree or two above freezing. It took 30 minutes of coaxing to empty the dredge. By the time he had finished, the shivering scientist had the coldest hand and forearm aboard ship.

When the dredge was empty, the net was pulled inside out. A hose, gushing sea water, washed the last remnants of specimens into a bucket. The full container and bucket were then moved forward to a position just behind a partially sheltering bulkhead, and the "washing" began.

It was two o'clock in the morning of April 28. The winchman had long ago secured his cable and the controls and had gone below. The *Chain* turned once more on her course, and except for the watch and the watch officer, everyone tried to get some sleep. All except the scientists who had been handling the dredge. An empty galvanized container was filled with sea water. It was fitted with an overflow spout near the top. A hose was pushed down to the bottom of the container and sea water swirled upward in the glare of the floodlight and ran out the overflow spout. Then, handful by cold handful, all the sand and rock and mud from the dredging net was placed in the water-filled can. The force of the swirling water floated the sand and particles of microscopic life to the surface and sent them out the overflow spout, where the entire discharge was caught in a fine wire sieve. The water promptly passed on through and ran into the scuppers, but all specimens lay revealed on the wire sieve. The large ones—the crabs, the eels and the fish—were taken by hand from the can of sand and mud and placed in plastic jugs filled with a preserving liquid.

The wind and sea rose further. The scientists' position in the lee of the bulkhead gave some protection, but not much. Waves began to break over the after rail. White foaming water sluiced toward the stern and disappeared overside. The wind was cold, down in the 30's, but the scientists grouped themselves around the sieve, collars turned up, staring at the flow of water that came endlessly from the spout. When a particularly interesting specimen plopped into view, there was a murmur of surprise and approval from the group. They seemed oblivious of the cold and the wind.

Cold sea water splashing through the sieve eventually numbed the hands of the man holding it. At the end of an hour he stood up stiffly and handed the sieve to a colleague. Then he went forward, clanging the bulkhead hatch behind him. Walking was becoming more difficult. The *Chain* rose and fell with a quartering roll in the heavy seas. A light burned in the officers' mess, but the warm cabin was empty. Everyone had turned in. The scientist got a heavy china coffee mug from the rack and drew a steaming hot cup of coffee from the urn. He stirred sugar into it and then sat heavily down, bracing himself against the roll of the ship. From time to time the portholes at which he gazed absently were white with the foam of a wave. At other times they were beneath the surface of the black sea as the *Chain* made her way toward the next station.

On deck the wash would continue as the scientists took turns having their hands frozen, until the last specimen had been caught and stored for examination later, back in the laboratory in Woods Hole. It took the whole night.

Dawn came and with it the storm.

3. ETERNAL WAVES

THE *Chain* was near the center of the atmospheric disturbance about which her captain had been warned long before. While he had admitted to the scientists that they would be having some "weather ahead," he had not told them of its severity, for a good reason. He himself did not know exactly how severe the weather would be. There was another, less logical reason for his taciturnity. If the storm were minor and he predicted a heavy one, he would look foolish. If the reverse were true, the same thing would happen. With caution born of experience, the captain said as little as possible to answer the queries that came his way.

When daylight came that morning, gray and sickly through a tangle of thick, sooty clouds, it was evident to everyone that the disturbance was not minor. The *Chain* altered course slightly to gain the most favorable position with respect to the oncoming wave pattern. It was a quartering course, sending the ship sidling up the forward slope of a wave and sidling down the rear slope with a minimum of roll or pitch. Even with this precaution, the *Chain*'s bow buried itself in the seas, white water drenched her foredeck, and the spray, lifted by the wind, streamed high, like smoke, to soak observers on the topmost deck. From that vantage point a few early risers watched the froth-white expanse of an ocean surface torn by the wind.

There was no single rhythm or pattern to the waves. Some medium waves appeared to be running counter to the large ones. Others moved across the large ones. The resulting upward spouting of many confluences made the sea a great carpet of maelstroms.

Hidden from the observers' view was another factor causing the turmoil. The *Chain* and the storm had met in the Gulf Stream, and this great body of water moving northward opposed the wind which was blowing nearly due south. The result was a white wilderness of ocean.

The watch officer had secured the open afterdeck. A line stretched across its only access points prevented anyone from wandering foolishly onto that dangerous area. The stern of the *Chain,* where the large A-frame was fixed, had no protective railing, and the railing itself was low so that it offered little help to anyone in danger of being washed overboard. As the bow buried itself into oncoming seas, the stern rose high above them. Then the stern would dip and a quartering rush of water would sweep over the low rail, blot out the deck with white foam and rush toward the open stern where water by the ton poured overside. Had anyone ventured there when one of the waves rolled aboard, he could have departed the premises in the flash of an eye.

Soon enough, scientists and crew adjusted to the unsteady motion of the ship, as countless mariners have done before them. The wrench and heave of the ship actually increased the feeling of safety for those behind the watertight bulkheads and hull plates. Since all areas of the ship were accessible through inside passageways, no one had to go on deck. Even so, the sea remained in sight. Through both luncheon and dinner, the roll of the ship and encounters with waves buried the portholes in dark water.

Viewed from the bridge, the sea that had made up was a confused sight. It might occur to an observer that here, at last, was one of nature's manifestations that man cannot reduce to a simple pattern for observation and analysis.

Perhaps not. But physical oceanographers are today making a good try at wave analysis. They know, for example, that the majority of waves on the surface of any body of water are caused by the action of wind on the surface. The moving air, great masses of it, has a tremendous amount of energy, and some of it, from that part of the air mass that comes in contact with the sea's surface, is transferred to the water itself. One present idea of how this is accomplished involves the fact that a wind, however steadily it may seem to blow, is constantly increasing and dying out, gusting and shifting direction. Such a passage of air over still water sets up a small series of ripples. Once these have been formed, the wind pushes against the sloping sides of the ripples, increasing their size and sending them hurrying faster forward. Soon a wave pattern develops and, in the open sea, the heights can grow to more than 25 feet with only a moderate wind blowing long enough and over a wide enough expanse of water.

The sea, of course, is never still. There are waves—or their aged relatives, swells—running always in some direction in the sea. There are also currents and tidal patterns alternately sweeping in one direction and then another. So a fresh wind from a new atmospheric disturbance will start a wave pattern that inevitably runs into some existing waves or swells. If a current is running counter to the main waves or across them, a complex combination of waves results.

In trying to analyze and account for ocean wave action far from shore, physical oceanographers approach the problem from two different directions. One is the traditional method of treating waves as a series of regular disturbances of some medium—in this case, water. From this viewpoint, waves can be described in terms of their height or amplitude, which is the vertical distance measured, in the case of water waves, from the crest of the wave to its trough. The term "wave length" refers to the horizontal distance between two succeeding wave crests (or other identical points on adjacent waves). These two dimensions of waves, when used in conjunction with the wave's speed through the water, can give oceanographers most of the information they need. They can, for example, tell how much force will be exerted by each square foot of wave surface should the wave strike an object.

But there are numbers of drawbacks to this kind of analysis. In an area where a sea is "making up," that is, where waves are being generated by winds, there frequently is no single wave pattern to be observed going in a particular direction. It is only after waves have left the so-called "generating area" and become undulating swells in a relatively calm sea that the above method of analyzing waves can be used with any accuracy.

Statistical methods of analyzing waves suit best when several wave patterns are intermingled. In such a case no wave is exactly like the one ahead of or behind it. And since the speed of waves depends on the ratio of a wave's height or amplitude to the depth of water through which it is passing, small waves overtake larger ones, combining their energy with that of the large wave for the moment of confluence. All of this may easily be occurring as swells from great distances far beyond the generating area enter the picture and confuse things even further. Physical oceanographers now know that the energy imparted to waves depends on three things: the duration of the generating wind, the distance over which the wind is blowing, and the speed of the wind. All of these affect the height and energy of the wave under analysis. By borrowing wave analysis techniques from communications experts, physical oceanographers have arrived at a means of predicting with some accuracy the probability of wave heights in any sea.

In a recent storm on the Atlantic Ocean, for example, the Italian luxury liner *Michelangelo* was proceeding toward New York. The ship was equipped with every modern device for safety, and the very size of the vessel gave both passengers and crew the illusion that the sea could cause them no harm. But a great wave struck the bow of the *Michelangelo*. It rolled over the tall bow of the ship, inundating the entire forward section. Steel plates were rolled back like tin foil from the bow. Windows in the bridge more than 80 feet above the water were shattered. The sudden collision of the ship with that giant wave threw passengers and crew to the decks. Three people were killed and many were injured.

The *Michelangelo* could have avoided that wave and the resulting loss of life and damage to the ship. Dr. Richard W. James of the Navy Oceanographic Office analyzed the storm—as his organization does every storm—and he used modern statistical techniques of analysis. The average wave height of that disturbance was approximately 30 feet and the *Michelangelo* was encountering those waves every twelve seconds. This means that the storm was not considered severe for a vessel the size of the Italian liner and it caused neither alarm nor concern among the crew or passengers. But according to Dr. James, if a thousand waves 30 feet high passing every twelve seconds went by the ship, there was one chance in twenty that a wave 2.2 times higher than the average would be encountered. Chance became reality that day in the storm-tossed Atlantic and a 66-foot wave did the damage. Naturally when it struck the hull, the crest moved over the deck and, striking the forward superstructure, flew higher still to crash against the bridge windows.

The office in which Dr. James works offers a special service to shipping companies. For between $100 and $200, a ship can be routed on a particular passage to avoid storms and the probability of meeting a wave like the one that struck the *Michelangelo*. Such routing has saved an average of 15 hours in each ship crossing, as well as untold thousands of dollars in savings to freighters whose cargo would have shifted and been damaged in such a storm.

For all the drama they create at sea, storm waves affect men more when they strike land. Much of the interest in waves is directed at learning exactly what occurs when large waves approach and hit a coastal area. This question, in fact, was so important during the landing of troops on foreign shores in World War II that it forced a quick study of the problem. A number of such invasions nearly became disasters for the Allies before a reasonable analysis was forthcoming. Fortunately the problem was solved with sufficient accuracy to prevent military failure.

Oceanographers have known for a long time that as a wave or swell approaches a shoreline, the bottom of the wave (the deepest point below the surface where water is disturbed by the wave's passing) begins to drag on the shoaling sea floor. When this occurs, the velocity of the entire wave form slows down but the motion of the water particles making up the wave increases at the top of the wave. At some point these particles will move faster than the wave form itself, and the familiar and often beautiful sight of a breaking wave will be created. As the breakers roll in endlessly on a shore, they bring with them energy gained from a wind perhaps thousands of miles away. This energy can exert tremendous forces, forces that erode the beaches, carve great sections of rocky cliffs away, and in stormy periods successfully wreck man's works along a shore. Such forces have been measured and are known to be as high as 6000 pounds exerted by every foot along the front of waves coming in. Thus, in a wave 100 feet long, 600,000 pounds of force are exerted on the shore when it strikes.

These heavy seas have been known to lift huge rocks weighing as much as three and a half tons and throw them over protective barriers 20 feet high. Smaller rocks have been hurled hundreds of feet into the air and have broken lighthouse windows on stormy coasts. In one reported instance a breakwater weighing 800 tons and securely fastened to the rocky sea floor with iron posts was moved from its position by the force of crashing waves. When men who witnessed that titanic push overcame their astonishment, they rebuilt the breakwater, using more concrete to bring the total amount to 2600 tons. A few years later, the seas once again rose and carried away the new, heavier structure.

A knowledgeable observer standing on a beach can tell much about a wave by watching and listening to the incoming pattern of disturbances. Of course if the shore area is in the wave-generating area, there is no question about the origin of the waves. They will arrive, one piling atop the other in an almost continuous roar of breaking surf. But if the air is calm and the sea relatively unruffled, long swells may approach from what appears to be a flat sea. The presence of the swells may only be noted when they have reached shallow water and become distorted into the familiar rounded wave shape. If as they approach the shore the tops rise and curl off, forming a wide and not very impressive whitecap, and if the breaking crest rolls easily down the forward slope of the swell, the wave came from a not too distant point on the sea's surface. But if there is some time lapse between breaking waves and if the swell approaches without a trace of a breaking crest until, near the shore, it rises swiftly and crashes in one final shore-thumping roar into its own trough, then we know that wave has traveled far across the ocean and perhaps from some giant storm thousands of miles away.

When waves leave the storm area, they can travel for surprising distances without apparent loss of energy. Swells can move through the open sea for as far as 2000 miles to expend their energy on a distant shore. Their average speed is about 15 miles per hour. They are always capable of being rebuilt into larger waves if they move into a second area where wind blows forcefully and over a period of time. There is one location, in fact, where continuous heavy seas dominate the oceans. This is in relatively high southern latitudes just north of the Antarctic Continent but below the land masses of South America and Africa. Here the prevailing westerly winds blow up huge seas, and there is an unobstructed circle around the earth to let the strong, far-traveling wind do its work. Waves in these latitudes frequently form a pattern of crests rising 40 feet above their troughs. Drake Passage, between the tip of South America and Palmer Peninsula, the northernmost reach of Antarctica, is probably more storm-tossed than any other area of the world's ocean. The famed Strait of Magellan lies here, and the history of the sea is filled with stories of shipwreck in this legendary area. No one knows how many vessels have been lost among the great waves of Drake Passage or in the savage rush of water against the rocky coasts of the southern tip of South America.

As Dr. James's statistical analysis has shown, average waves 40 feet high can sometimes combine to produce individual waves of terrifying dimensions. And if the average is higher—as it is during storms over that region—the occasional monster waves are even taller. Waves of more than a hundred feet in height have been reported by experienced seamen.

In one case, the officer on the bridge had the opportunity actually to measure the height of one of these astonishing waves. The incident occurred during the passage of a U.S. Navy vessel in a storm far out on the Pacific Ocean. The watch officer on the bridge glanced astern just as one of the giant waves formed and bore down on the ship. He noted that its crest rose above a particular point on the mainmast above the crow's nest. The ship was on an even keel in the preceding trough and when the wave and the excitement had passed, simple triangulation showed the height of that wave to be 112 feet.

The most severe storms at sea that stir the ocean and produce waves are hurricanes or typhoons. Such storms are termed hurricanes in the Atlantic but have retained their traditional name of typhoon in the western Pacific. These storms are the result of a low-pressure atmospheric area building over the ocean. Like a funnel, this low pressure invites air on its edges to pour in from the surrounding high-pressure air masses. As the air flows in, it describes a circular motion which is counterclockwise when viewed from above in the Northern Hemisphere. When the low pressure intensifies and more air is poured into the system, the winds along the surface of the ocean increase, and the size of the storm increases until, in the largest of such storms, the diameter of the circle of disturbance may be as great as 500 miles. The circular winds have been known to have velocities of 150 miles per hour or slightly more (frequently the wind indicator—anemometer—blows away when such velocities are reached). The entire disturbance may move at a forward speed of anywhere from two or three miles an hour to around 50 miles per hour. Sometimes the storm speeds up and then, for no apparent reason, slows down again. This makes the time of the storm's arrival at any point difficult to predict.

But the waves set up beneath the storm leave the storm area and move far across the sea, sometimes giving advance warning to men on shore of the approach of a hurricane. Indeed in many sparsely inhabited places on the earth, this arrival of storm swells is the only warning the coastal inhabitants may have of the hurricane until the low hurrying clouds make their appearance over the horizon. There is a special situation that heralds great danger for human beings in these storm waves and the accompanying winds. For the waves continue to build as the winds continue to blow. If such a storm system moves toward a coast, the part of the great, turning circle of the storm that is driving directly onshore can not only build great waves but can actually pile up the water during the times of the storm's greatest intensity. Should this occur in a large bay or other indentation on the shore from which the waters cannot escape, all the elements of a tragic coastal flood are present except one: the proper tidal action at that precise time and place.

If the storm strikes, piling up water and still building waves as the tide is flowing into the bay or estuary, then tremendous damage is done to all of man's work along the shore and often to man himself unless he is warned and can flee.

Such situations have risen many times. In Galveston, Texas, in 1900 a storm struck the town and flooding began. For over four hours the storm drove water into the streets from the bay. And then a sudden surge appeared and the water rose four feet in an instant. On top of the already flooded streets this spelled disaster, and 600 people were drowned.

An even greater story of disaster from waves striking the shore occurred in the Bay of Bengal in 1876. With the storm piling water up offshore on the low-lying coast, the tide was ebbing. Had the storm passed over during that time perhaps fewer people would have died. But the storm stayed almost in place until the tide turned and began coming in. A great storm wave rode the tide to the shoreline, covering everything far inland to a depth of 40 feet. Although it was impossible to count the dead, estimates of 100,000 drowned were considered reasonable.

Such surface waves as the storms at sea create are among the least influential of all movements of the sea. Chief source of water motion in the sea is the tidal influence of the moon and the sun. The total effect of the gravitational pull of these other-than-earthly bodies is far beyond comprehension or reasonable calculation. Surface waves disturb water to no great depth in the ocean. But the tidal pull of the moon affects all the water to the ocean's deepest depths. It also has a measurable effect on the surface of the earth, but since the surface is a relatively rigid solid that resists movement, we have little visible evidence of this action of the moon and sun. Water, being a liquid, can flow under the influence of any force, and flow it does as the moon swings around the earth while the earth in turn swings around the sun.

In theory the tides should be regular as clockwork—the clockwork of our solar system. And if the earth were completely covered by water and the ocean basins formed a smooth sphere with no undersea mounts and no shoaling floor, an observer from a satellite could view the perfect rhythm of tidal motion. This symmetrical state of affairs does not exist, however. Continents rise above the sea. Shorelines are irregular. There are indentations, bays, estuaries. On some parts of coastlines, there is a wide shallow continental shelf while on others there is none. Some coastlines run east and west, others north and south. All these discrepancies plus a number of others upset the pure theory of tidal movements.

One of the most important of other effects is a sort of natural swinging motion set up when water is disturbed in a confined basin. If you drag your hand across the surface of the water in a pail or basin or bathtub, a wave rises on the side of the pail or tub, reverses itself and goes to the other side, where it rises, is reflected and again returns to the first side. This motion soon takes on a curious appearance as it repeats itself. The center of the water surface is moved less and less while the alternate rising and falling of the water on the sides continues. It becomes much like a children's seesaw. The center where the plank is hinged does not move, but both ends of the plank go through the greatest up and down motions.

Such a standing wave has the special name of seiche. In the case of tides, the ocean basins themselves act as the pail or tub. The disturbing force is the gravitational attraction of the sun and moon, which produce a regular tidal sweep. But the seiches set up by the tidal motion tend to reinforce the tides as they sweep onto the continental shores, for here both tidal action and the seiche action can reinforce one another.

In the case of the North Atlantic Ocean basin they do just this. For the time between successive rises of seiches depends on the physical dimensions of the basin. This time between similar positions of the standing wave is called the period of vibration. In the case of the North Atlantic basin that period is approximately 12 hours—almost exactly the period of tidal motion. The result in this basin then is the reinforcement of tidal action on the shores. This is particularly noticeable in the famous Bay of Fundy where tides of 40 and 50 feet are expected twice each day. If we could measure the tidal rise and fall in mid-Atlantic we would find it nearly imperceptible. There the effect of the seiche is zero, since that midpoint is exactly where the hinge would be on the seesaw.

Oceanographers do not have a way of measuring tides accurately in the open ocean. But on the island of Martha's Vineyard just offshore from Woods Hole, the south beach has a moderate tide range—not more than three feet. A check of the tide ranges on shore points going northward toward the Bay of Fundy shows a regular increase. On the coast of Maine, ten- and twelve-foot tides are common and these increase until the mightiest tides in the world can be observed at the head of the Bay of Fundy.

With complete indifference toward all of man's neat

theories, tidal movements present a complex picture from coast to coast and from point to point along the same coast. And there are places where our information about tidal movement is still incomplete. Small bays, shore currents, winds, the slope of the sea bottom offshore can all affect the time of high tides despite our theories. The only way actually to tell when high tide will occur at a particular point on a particular coast is to plant a tide gauge there and measure the coming and going of the tides.

It is important that man know about the tides. Men who make their living from the sea govern their lives by the coming and going of the tides. Many lobstermen, for example, living on offshore islands along the coast of Maine are almost total captives of the tide. Their dories may be moored where they are afloat only near the top of the tide. Their large fishing boats may not be able to pass through critical channels in times other than high water. So they learn early in life to plan their days in response to the tides. Indeed all mariners everywhere must obey the tides, for even man's largest luxury liners cannot dock safely except at the time between great tidal currents flowing in the harbor. They hurry or wait to reach the dock at slack water, between the times of most extreme tidal flow.

In the history of life on the globe, the tidal areas of the shore were vitally important. For it is a zone that is sometimes covered with water where sea creatures can live and sometimes exposed to the atmosphere where land-dwelling creatures breathe air. There are some creatures and plants that today live part of the time out of the water and part of the time under water. From the standpoint of evolution, the tidal zone was therefore extremely critical, for it nurtured those forms of life that eventually were to adapt to a continuous life out of the water. Both plants and animals of the land today are forms of life that originated in the sea and made their eons-long journey from water to air through the transition zone alternately covered and exposed by the tides.

INTERNAL WAVES

The physical oceanographer who classifies wave forms of the ocean must consider one type not observed by most of us. These are called submarine or internal waves. Since they are not immediately evident to the observer, it is difficult to picture them and understand their effect on the waters and life in the sea.

Oceanographers know that bodies of water having different temperatures, densities or salinities have a tendency not to mix. Scientists can, in fact, frequently identify the ocean or portion of it that a sample of water comes from by the saltiness of the water. But with great masses of water being separated from adjacent masses by one or more of these factors, there is a resulting boundary area or interface, as it is called. Internal waves can be set up by many sources that move along these interfaces just as surface waves move along the boundary or interface of the atmosphere and the sea itself.

Such waves contain great energy and many of them are larger than the surface waves. Some have an amplitude of up to 300 feet. These giant submarine waves travel unseen throughout the oceans of the world. There is little question but that these waves, which are now under study by physical oceanographers, must affect the lives of millions of sea creatures. Schools of herring are thought to be driven into the bays of coastal European countries by the force of these waves breaking far beneath the surface over the edges of the continental shelf.

To grasp the idea of interfaces and the internal waves that travel along them, a simple experiment will help. If water is placed in a long glass-sided container like a home aquarium, and kerosene or other oil poured on top of it, the boundary area or interface between the oil and water becomes visible. If the water in the container is disturbed at one end to the depth of the interface, a surface wave will be set up, of course, on top of the kerosene. Observation of the interface will reveal a wave pattern running along it—a true internal wave.

The only misconception you may get from this small demonstration is that while the interface you observed was horizontal, interfaces within the ocean are not necessarily horizontal at all. Masses of water having different densities, for example, may meet one another on any sort of boundary shape and incline to the horizontal at a considerable angle. This, in fact, is the rule rather than the exception. Such an interface may be curved and tipped. It may even be almost vertical and wavy like a curtain. But along it run the secret internal waves.

Should a submarine cross such an interface when a large wave is passing along it, the undersea vehicle will experience the rolling and pitching that any surface vessel does when it encounters such waves as the *Chain* did in its stormy passage across the Gulf Stream. For the large undersea waves can be thought of as acting much like surface waves. When they run into an obstacle, they become huge silent breakers where tons of water, making no noise whatever, are turbulent and dangerous as any surf. Such an obstacle may be solid like a sea mountain or cliff, or it may be a strong deep current like the Gulf Stream that interrupts the passage of such a wave.

There are only two bits of evidence of internal waves that ordinarily come to the attention of men on the surface of the sea. One of these is an unnaturally smooth, slick area or long wavy one of surface water where no surface waves form for a long time. It is believed this is surface evidence that the energy that would normally form or support a surface wave is being used rather to create an internal wave running along an inclined interface that comes to the surface along the slick.

The other evidence is of more importance, particularly

to ships in polar regions. For in either the Arctic or Antarctic, near a field of ice or near the face of a glacier or Antarctic ice shelf, a layer of relatively fresh water from melting ice will form. Between this layer of fresh water and the sea water beneath it lies a nearly horizontal interface. Should the layer of fresh water be approximately as thick as the ship's propeller is deep in the water, the ship will have great difficulty maneuvering. For the energy of the turning propeller does not drive the ship forward as it is supposed to do. The energy is all used up in setting up internal waves along the interface. This is such a common occurrence in all parts of the sea (although most common in polar waters) that seamen refer to such a stretch of ocean as "dead water," and so far as they are concerned, they are absolutely right.

TSUNAMI

Men tend to focus their attention on what is observable directly, and particularly on natural phenomena that cause them distress and represent a danger to their activities. This being the case, it should not be surprising that a great deal of study has gone into the causes and structure of the most terrifying of the ocean's many manifestations: the tidal wave. The term itself is not accurate, since waves of this type are not caused by the tide. The more acceptable terms used today to identify what have been called tidal waves are seismic sea wave and tsunami—a word of Japanese origin. These are the waves that occasionally appear on many coasts throughout the world. They come at most in a short series of two or three. They flood the coastline quickly and often with terrible destruction to man and his works.

The source of such waves does not even lie in the sea itself. Tsunami occur whenever there is a shifting of the earth's crust beneath the sea or when a great submarine landslide takes place on the precipitous continental slope. The most common source of tsunami is in the Pacific Ocean—which is far from peaceful in an oceanographer's understanding of it.

The floor of the Pacific Ocean basin, particularly in the western edge, is marked by a series of trenches. These canyons in the ocean floor are the deepest penetrations toward the underlying mantle of the earth, and it is here that many undersea earthquakes occur. These earthquakes are classical in their action, with some long strip of the earth's crust either rising or dropping or being displaced, horizontally. When this occurs beneath the sea, the earth being displaced naturally displaces an equal volume of sea. Such a happening sends out tremendous amounts of energy—as a loud explosion in the air sends out a sharp, deafening wave of sound and energy. If such an earthquake takes place on the sea floor, an expanding spherical wave is set up. Like a balloon being blown up under water, it grows and rises until it breaks the surface of the sea. If the earthquake caused a lifting of the earth's crust, then the upward displacement of water boils to the surface, raising the sea as the bubble of energy breaks through. This sudden elevation of the sea surface over many square miles is not physically high—perhaps no more than a few feet. But the volume of water is immense. This sudden displacement of sea water sets up the astonishing surface wave that is called a tsunami.

The other principal cause of tsunami comes from undersea volcanic action. Such break-throughs in the earth's crust can occur on the floor of the sea as easily as above the ocean's surface. When such volcanic action spews molten lava from a vent far beneath the sea, the heat turns the surrounding water to steam—and when water is converted to steam its volume increases tremendously and in an instant. Two forces suddenly confront each other.

One is the force exerted by the expanding steam. The other is the weight of the tons of water on top of the undersea volcano. The steam wins in this struggle, and there is an upsurge of water, the creation of another seismic sea wave or tsunami. Unlike other surface waves, the speed of this one in spreading is unbelievably great. Speeds as high as 500 miles per hour have been reported. The wave, if it occurs in the open ocean, is circular in shape and it travels far and fast. As the water subsides, it oscillates up and down a few times before dissipating all of its energy. Thus several more waves of varying intensity follow outward on the heels of the first one.

Should the changing of the earth's crust be in a downward direction, a similar situation results, but the initial surface wave is set up not by an upthrust of water but rather by a sudden falling or lowering of a vast area of the sea.

To a mariner in the open sea, such waves may not be noticeable at all. Indeed these waves do not damage anything until they near the shore of some unfortunate island or continent. Then, if there are trained observers to observe the phenomena, the first indication that a tsunami is about to strike is the unusual withdrawal of water from the shore. The water may recede far beyond the lowest low tide. It begins quietly enough. Bathers on such a shore may not be aware of the gradual recession of the water until extraordinary portions of the sea bed off the beach are exposed. But when this abnormal withdrawal happens, the knowledgeable observer sounds the alarm. Then, with whatever time is left to them, the people flee inward to higher ground. There is not time to do more than turn and run, for within minutes the tsunami will be upon them.

Some people who have lived through such catastrophes describe the incoming water as making no sound other than a strange hissing effect at first. But as the wave approaches the beach it acts like any other surf wave breaking on a shore, only magnified many times. The great wave can in some cases pile up before it breaks to heights of nearly 100 feet. And when it does break, with a roar of terrifying loudness, it sweeps everything before it. It can take ships and move them inland hundreds of feet. Breakwaters, docks, waterfront buildings and small ships are obliterated. If warning is not given and people have no time to flee, lives are lost by the scores and hundreds.

The many islands of the Pacific as well as the coast of Japan are vulnerable to tsunami, and many of them have been struck so frequently that there are formal warning systems to alert people on the beach or near the shore. At Hilo on the island of Hawaii, for example, a siren wails when the first warning of a tsunami is noticed: the quiet and dreadful withdrawal of the water. In recent years seismologists and oceanographers, who continuously observe the earth's crustal movements and the unusual ocean effects that follow, have begun to broadcast tsunami warnings by radio to all populated islands and shore areas of continents bordering the Pacific Ocean. Thus an undersea earthquake in the Aleutian Trench will make seismographs around the Pacific fluctuate wildly. Within minutes the radio warning will go out to stations around the ocean. Since the speed of the tsunami is known, it is a simple matter to tell a station in Hawaii, for example, when to expect the wave's arrival.

One of the most destructive tsunami on record occurred on the coast of Japan on the island of Honshu in 1896. It was a warm summer day and the villagers had flocked to the seashore to escape the heat. During the afternoon of June 15, the bathers felt a small earth tremor which did not disturb them at all. That section of the world frequently is joggled by earthquakes, as are all other lands encircling the Pacific. This one was so mild that no one paid any attention to it. From experience the people knew it was an undersea quake somewhere offshore. In fact it occurred in Tuscarora Trench lying off the coast of Japan.

Shortly after the quake was felt, the waters along the shore began to recede. At first it was exactly like the tide going out—and no one thought much of it. But the water continued to withdraw. More and more of the ocean floor off the beach lay exposed to the sunlight. Parts of it were exposed for the first time in anyone's memory. Still the people remained on the beach. The first inkling they had that catastrophe was about to strike came by way of sound. It has been described as first a faint whisper, like the sound of rain falling on the surface of water. But the sound grew as the water began to return. It increased to a deafening roar as the great tsunami began to encroach on the land. The wave front piled up exactly like a giant breaker—only larger than any that men have ever seen.

In some places the height of that single wave reached an estimated 100 feet. Then it broke and swept inland taking with it everything in its path—including 28,000 lives.

4. A PROFILE OF THE SEA

THE *Chain* labored over the rough seas of the storm and finally passed through the Gulf Stream. But she did not immediately find smooth sailing on the other side. The wave pattern persisted, the wind still blew at gale force and the bow, rising and falling with monotonous regularity, frequently took white water over the foredeck. The ship rolled, too. In the officers' mess, those who made it to meals had to brace themselves with elbows, hands and feet. A sudden lurch of the ship would send bottles of ketchup, mustard, the sugar bowls and salt and pepper shakers sliding from one side of the table to the other.

At lunch on the second day out several scientists and ship's officers dined in silence as each concentrated on the problem of reaching for skidding plates and made valiant efforts not to stab themselves with their forks as the ship continued its erratic motion. Then, due to a slight error in helmsmanship, the bow of the *Chain* turned a little too far away from the oncoming wave—which happened to be one of the larger variety. The ship started to roll, and even the greenest landlubber knew his effort to retrieve bottles and plates was doomed.

Moving almost broadside to the wave, the ship lay over on her starboard beam and then began to return. It was not a graceful scene. All glassware crashed to the floor. Silver skidded the width of the mess cabin. The cook wailed from the galley and the messboy delivered himself of several choice phrases. If anything, the return roll was more spectacular than the initial heeling over. Portholes buried themselves beneath the raging green water. The force of it pinned people in place and the hapless utensils flew the opposite way across the floor. Trying to retrieve a few objects, the messboy footed his way through a mild surf of spilled ketchup and mustard and, losing his toehold, began to skate across the deck. He grabbed a bolted-down table and hung on. He had exhausted his vocabulary. Once more the ship rolled before the green helmsman on the bridge was replaced and a more experienced hand took the wheel. The exciting seizure was over.

When order was restored and the spilled food and

broken plates and glasses swept up, scientists made their way to cabin or laboratory to prepare for the next station and the work ahead. The storm had forced the ship to pass by one station on this outward leg of the trip. With seas that rough, no equipment could be put overside. The chief scientist sent word to the bridge to proceed to the third station in the hope he could find quieter waters there. So the ship's captain continued on his course with as much speed as seemed advisable in rough water.

At midnight on April 28, the bridge reported "On station" at 37° 00′ North Latitude, 69° 00′ West Longitude. Depth: 5000 meters. Beneath the *Chain* lay the floor of the North Atlantic basin, an all but unknown region of the earth.

The water still was rough, although the waves were subsiding. In the hope that calmer waters would soon surround the ship, the senior scientist suggested that the deep dredge be postponed until other work had been completed.

Accordingly the scientist in charge of the hydrocast soundings readied his gear. For gathering his data, the scientist had to make use of the winch fixed to the starboard side of the ship. Like its larger counterpart, the drum of this winch had enough steel cable to reach to the greatest depth in the ocean. The only difference lay in the weight of cable. Both winch and cable were much lighter and smaller than the dredging winch. The scientist attached a heavy lead weight to the end of the cable and signaled the winchman to lower it over the side to a depth of 200 meters. At that depth, the winch stopped and the scientist emerged from his laboratory carrying a long metallic cylinder on the outside of which thermometers were fastened. Each end of the cylinder had a clamp which could be tightened by a thumbscrew. The scientist carefully fastened both ends of the cylinder to the cable. When he was sure it was firmly in place, he motioned the winchman to run out another 200 meters of cable and he then attached another cylinder exactly like the first. All told, there were 25 cylinders to attach.

The cylinders bore the name, well known to oceanographers, of Nansen bottles. They are a nearly foolproof device used to gather water samples and measure temperatures at different depths beneath the sea.

In principle they are simple to operate and it is in this simplicity that much of their value lies. For instruments lowered to any great depth in the open sea have a peculiar habit of not working properly. At least not as the surface living animal called man had expected them to work. For the sea is a new and strange environment. We do not know precisely what conditions will be met with and cannot tell what keeps instruments from functioning.

The Nansen bottle has stood the test of time. It was invented at the turn of the century by the great Arctic explorer and oceanographer Fridtjof Nansen. When the bottles are in place at their desired levels on the cable, a metal "messenger" is fitted to the upper end of the cable by the scientist. He releases it and it slides down along the cable. The clamps are fixed so they do not stop the messenger, and as it slides by each bottle it releases the upper clamp. Pulled by gravity, the bottle begins to "fall" outward and downward through a semicircular arc, hinged at the lower clamp level. While this motion is happening the valves are opened by the turning, admitting samples of water at each depth. By the time the bottle is hanging upside down, the valves are closed again and the sample is trapped and ready to be hauled to the surface. At this time too, the temperature of the water at the various depths is taken by two so-called "reversing thermometers." These thermometers have a mercury column that is arranged to "break" during the turning motion of the bottle and thus preserve the temperature at the depth of the bottle. Two such thermometers are used. One of them is protected from the pressure of the water while the other is exposed to it. When the bottle is brought to the surface, there is a difference in the readings of the thermometers. By correct interpretation of this difference the scientists can tell what the water pressure was at the depth of the bottle.

As the winchman slowly hauls up the cable, the scientist removes each inverted bottle when it surfaces and takes it into his laboratory. And when the last bottle is removed, he struggles with the ballast weight on the end of the cable, swings it over the rail with considerable grunting and brings it to rest in an iron stand by the starboard rail.

The bottles in the laboratory now attract the scientist. He has gotten what is known as a profile of sea water—in this case reaching down 5000 meters to the sea floor. By extracting samples of the trapped sea water from each bottle he can analyze it for any of several items that may be of interest to the particular project. Among the things that an oceanographer would like to know are the free oxygen content of the water, amount of nitrates, inorganic carbonates, phosphates; amounts of other minerals or individual elements, amounts of dead organic matter from the uncountable billions of small marine creatures that have died and are slowly drifting to the ocean floor to contribute to the increasing layer of sediment that covers the ocean bottom.

In addition, he can discover how the temperature and pressure vary all the way down, as well as the salinity of the water. This profile, at this station, becomes one of thousands of other such profiles being recorded daily and yearly by oceanographers all over the world.

With just the temperature and salinity information, the oceanographer has been able to learn a great deal about the seas and the movements of masses of water the world around. Although these modern measurements were begun not many years ago, scientists have a fair knowledge of the major motions of sea-water masses. For each of the factors—temperature and salinity—affect the density of sea water. That is, given a particular volume of water, say a cubic foot, a sample of cold water will weigh more than a sample of warm water. Also a sample of water with dissolved salt in it will weigh more than a sample of fresh water. If these samples were carefully released in a great body of water, the heavier samples would sink below the lighter samples. And they would sink in the ocean until they reached a depth where the density was exactly the same as each sample's density. They would, in fact, "find their own level."

Curiously, large masses of water having the same salinity and temperature tend to keep these characteristics for a long time. Oceanographers can identify the source of much sea water by measuring its salinity and temperature—and they can measure these characteristics to a high degree of accuracy. A change in salinity of as little as four parts in 10,000 can have meaning in the data charts of today. Temperatures can be measured to hundredths of degrees and even those small changes have meaning.

One of the clearest examples of the motion and character of these large masses of ocean comes from the antarctic regions. It is here, around the edge of the frozen continent, that the coldest water in the ocean is formed. During the deep long winters of the antarctic night the coldest water sinks. It flows along the continental shelf surrounding the continent and slides like a fantastically large Niagara in slow motion down the continental slope to the sea floor. At the bottom of the sea around the continent the water has a salinity of 34.62% and a temperature of minus 1.9° C. It is not frozen, because only pure water freezes at 0° C. With the "impurity" of salt in it, it remains a liquid below zero. This water forms what oceanographers call antarctic bottom water. It moves along the floor, pushed by the cold water behind it. It has been detected in the North Atlantic Ocean.

The surface water around the Antarctic Continent moves northward until it reaches a point about 55° South Latitude. It still is cold water by any standards, but as it moves northward its temperature rises slightly. At 55° South Latitude a marked rise in surface temperature has been noted by oceanographers. This line, circling the continent and moving seasonally both northward and southward, is called the antarctic convergence. It is well marked by the sudden rise in temperature and this is where the warm subtropical water of the South Atlantic is met. The colder water, being more dense than the warmer water, tends to sink. It still continues to flow northward beneath the surface for a distance, and then it reverses itself and doubles back, heading southward at considerable depth. This water is termed antarctic intermediate water. It has the lowest salinity of any southern water. This particular water type is easily detected all around the globe by its characteristic minimum salinity except in one location: the Indian Ocean. Here, water of nearly the same density is met flowing southward from the Red Sea. It erases the character of the antarctic intermediate water almost completely.

Arctic water in the Northern Hemisphere finds its way toward the equator but not in such large quantities as antarctic water in the south. The reason for this is the arrangement of land masses on the globe. Open seas to great depths surround Antarctica, but the Arctic Ocean is surrounded by land masses, and there are only two possible escapes for the colder water of that polar region. One is over the narrow and shallow sill running beneath the Bering Strait in the North Pacific. The other is over a wider but still rather shallow sill running from Greenland to the coast of northern Europe. Consequently most of what is termed arctic water is formed south of those sills.

In those lower latitudes, winter temperatures are not so severe and the final minimum temperature of arctic water does not match its southern counterpart. It ranges from 2.8° C. to 3.3° C. and is formed primarily in the Labrador basin and off the southeast coast of Greenland. As might be expected, the less cold water is less dense and although its salinity is slightly above that of antarctic bottom water, it does not sink to the very bottom in its southward journey. This water travels the entire length of the Atlantic Ocean from north to south until it finds itself flowing slowly over the antarctic bottom water in the Southern Hemisphere. Finally, at about 50° South Latitude, it is sandwiched between antarctic bottom and intermediate water. When it reaches a higher southern latitude, it rises to the surface, pushed upward by the slope of the falling bottom water.

This vast upwelling of water somewhere near 60° South Latitude brings with it the highest concentration of plankton of any water in the world. Mariners in those regions report floe ice heavily encrusted with plankton life. And in some places the sea water itself has been described as having a mushlike consistency because of the concentration of plankton. If ever there was a crop being grown by the world ocean, this is its most fertile field. As might be expected the larger, higher forms of animal life abound here because plankton is a primary source of food for small fish as well as whales.

On the surface, at the antarctic convergence, the warm water of the South Atlantic is turned under and back by the sinking cold water. It begins, at medium depths, to return to the warmer tropic areas from which it came. It is cooled by its southern journey and so sinks below the surface for the return trip.

In the subtropics of the North Atlantic, the warm waters north of the Equator flow northward until they meet a less well-defined line of cold water formed in the subarctic waters. But here, in the North Atlantic off the coast of Europe, another mass of water acts to alter the vast oceanic circulation. It is warm, very saline water from the Mediterranean flowing through the Strait of Gibraltar. This water's density is such that it sinks to a depth of about 1000 meters and mixes gradually with the Atlantic water in that area.

The thousands of cubic miles of South Pacific water behave much like the South Atlantic water. There is antarctic bottom water moving northward, antarctic intermediate water of minimum salinity and overlaid by southflowing water of the southern Pacific tropical regions. But the northern Pacific circulation is considerably different. Arctic water from the north polar basin is almost totally absent because of the shallow Bering Strait region. Thus only subarctic water from the Bering Sea and the Okhotsk Sea north of Japan is available to form bottom water.

None of these great masses of water that are in motion on the face of the earth moves fast. They are by no means regarded as currents since their velocities may be only three or four centimeters per second. But they have a definite effect on the climate and the weather as well as on the prevalence of sea life because, while they move slowly, there are vast amounts involved.

Where water is cold, it can contain more dissolved gas. Polar surface water has the highest percentage of oxygen of any oceanic water. When it sinks, the oxygen is trapped in that body of water and can support life at depths we believed until a few years ago could not contain life of any kind.

CURRENTS

Unlike the great ocean water circulation, currents are, at least on the surface, more easily charted and understood. They also are comparatively smaller and more mobile so that their motion has been observed for centuries, and acceptable charts of the most evident surface currents have long been in use. Indeed if there is a distinction between ocean circulation and ocean current, it is the speed of motion as well as the volume of water transported.

Currents are formed primarily because of the earth's rotation, the winds that blow across the surface of the sea, and the effects of the land masses forming the ocean perimeters.

A study of prevailing wind patterns shows that the several wide belts of winds surrounding the earth set up corresponding surface currents. North of the equator, for example, there is a region where the winds blow constantly across the sea's surface from northeast to southwest. These are known as the northeast trade winds and, if the earth were not rotating, would blow directly from north to south. The earth's turning, however, gives rise to an effective force that diverts winds (or any other moving object like a bullet, ocean current or projectile) toward the right in the northern hemisphere. Hence these winds begin a right turn and continue to increase it so that in the western reaches of the Atlantic just north of the equator, the "turning" of the winds has brought them gradually to an almost east-west position. These trade winds account for one of the strongest currents known: the North Equatorial Current, which flows from east to west above the equator until the current is scattered by the Greater and Lesser Antilles Islands. The masses of water moved by the current eventually flow into the Gulf of Mexico and, if it were not for the Central American land link between the two continents of the western hemisphere, this current would continue into the Pacific. It is, however, turned back. While it is in the Gulf of Mexico it receives more heat from the sun and emerges around the tip of Florida as the Gulf Stream.

The Gulf Stream is one of the best known and most influential currents in the world's oceans. Certainly it is of primary importance to all people in the eastern United States and in Europe, for it modifies and to some extent controls the climates of those regions.

The word "stream" does not accurately describe this huge river of the sea. In terms of oceanic water masses, it is relatively small in cross section but it is by no means a mere "stream." As it begins its long and now well-plotted course, it is indeed a mighty river in the sea, measuring over 90 miles wide and a mile deep, proceeding at a rate of about three knots. It is, in effect, a huge moving heat-transporting system. For its temperature is raised in the Gulf of Mexico and, with the high heat-retention character of water, it holds that heat, parceling it out slowly over its long journey, which does not truly end until its waters finally are cooled in the Arctic Ocean near the island of Novaya Zemlya. The tracing of this great oceanic river—that winds and bends and occasionally splits into small tributaries—is worth all the effort. Oceanographers and mariners before them have studied it carefully ever since the first chart of its energy-transporting course was made at the insistence of Benjamin Franklin around the year 1770. Navigators of the early sixteenth century had run into this stream of water and frequently, in the sailing vessels of that day, had been unable to make any headway against the stream even though they had favorable winds. But soon the mariners coming to the New World learned that they could find good sailing westward on the northern equatorial current. They could then sail the Gulf Stream north and east for a time until they reached the Middle Atlantic area of the east coast of North America and then strike out on an eastward course across the broad Atlantic, homeward bound.

Franklin's chart was sent to shipowners in England, and the untiring Franklin suggested they could take advantage of the stream if they knew of its existence and paid attention to the chart. According to some reports, the shipowners paid little attention to Franklin's helpful hint. Yet even today, with powered ships and steel hulls, ship captains use and respect the potent Gulf Stream current. They always take advantage of it to reduce the time of their voyages.

It is not as an aid to navigation, however, that the Gulf Stream gains its importance. It is as a modifier of climate. When the warm waters leave the gulf itself and move northward along the coast, the effect of the Coriolis force sends the stream angling a little farther from the shores of the southeastern United States. The coastline there, as a matter of fact, has four major land forms that were created and are now maintained by the Gulf Stream. In passing near shore along these southern states, large eddies of water developed that carried sand from one part of the coast and deposited it in another. Thus Cape Hatteras, Cape Lookout, Cape Fear and Cape Kennedy were born.

Northward and eastward the Gulf Stream moves, bearing always to the east. At about the latitude of Cape Cod, Massachusetts, it is well out to sea but a minor current has split off of it and moved north to modify the south-flowing cold water of the Labrador Current. The strangely temperate climate of parts of Nova Scotia is due to these warm waters from the Gulf Stream. South and east of Greenland other currents break away from the major stream and warm, in their turn, the southwest coast of Greenland and the southern shores of Iceland.

In these latitudes the Gulf Stream is moving almost directly east in response to a belt of winds from the west, to the configuration of the ocean floor near the continental shelf off Newfoundland and the ever-present Coriolis force. As the Gulf Stream approaches the coast of Europe, it splits into two major parts. One of them is named the Canary Current and it washes the shores of Spain. Some of this water finds its way into the Mediterranean Sea through the Strait of Gibraltar. The rest of it goes farther south along the northwest shoulder of Africa. Here the water completes its great clockwise gyration around the Atlantic basin as the Canary Current merges once more with the North Equatorial Current to repeat the endless cycle.

The water of the Gulf Stream that turned north as the European coast was neared moves along the Scandinavian coast and is renamed the Norway Current. The heat from the Gulf Stream tempers the climate of Britain and of the Scandinavian coastline. Even the Norway Current contains some usable and valuable heat energy. It rounds North Cape into the arctic basin and brings enough warmth to make the ports on the northern shores of Russia usable throughout most of the year while Baltic Sea ports far to the south remain locked in ice. Actual detailed studies of the Gulf Stream and other North Atlantic currents show more complex patterns, but the main climatological effects result from the larger general flow, as explained above.

The pattern of the Gulf Stream in the North Atlantic and of all the surface currents there that complete the principal gyration around the borders of the continents is repeated in the South Atlantic, with the important difference that the effect of the Coriolis force is to send the great gyration in a counterclockwise manner, driving the currents to the left as they proceed rather than the right. Thus the South Equatorial Current flows from east to west below the equator and most of it turns southward at the easternmost shoulder of the South American Continent. As it flows south along the coast of Brazil it takes the name of the Brazil Current, and when it meets the antarctic drift moving from west to east it turns eastward itself, bending slowly northward finally along the southern coast of Africa until completing a cycle as it approaches the equator once more.

The longest uninterrupted current flowing in the world lies in the Pacific where the North Equatorial and South Equatorial currents flow from the western shores of South America all the way to the Indies. In both the North and South Pacific Oceans there is a repetition of the great gyres found in the Atlantic. Only in this greatest ocean on the planet the circular pattern is larger.

Knowledge of these surface currents produced by winds and other forces is not yet complete, but as more and more oceanographic expeditions are sent out, more information can be gathered and fitted into the store of past knowledge. A good example of increased activity in analyzing ocean currents and water movement is the recent International Indian Ocean Expedition when 15 nations sent ships and oceanographers into those little-known waters for five years of consecutive research effort. This ocean is unique among those in the rest of the world—not only because it had not been studied. As has already been pointed out, it is in the Indian Ocean that antarctic intermediate water vanishes at its usual location because of Red Sea water. Another unique aspect of the Indian Ocean is that it supplies more visibly and more directly rains for the great subcontinent of India by means of the seasonal monsoons. With the turn of the seasons, the winds begin to blow northward across the subtropical waters, bringing millions of tons of water vapor in the atmosphere that sweeps across the inland areas of India and Southeast Asia. The moisture-laden air, rising across the land, cools and the water vapor condenses to produce the torrential monsoon rains that make the land fertile. If this rain does not come, crops, animals and men will die. Such a situation nearly occurred in 1966. With the monsoons delayed for some unknown reason, India's millions faced drought, and some cities were on the verge of evacuating people to other areas. But the rain finally arrived and those extreme measures were not necessary.

This monsoon phenomenon of the Indian Ocean and the land masses to the north of it gives rise to one of the most newly discovered currents in the world: the Somali Current, as it has been named. It is unique among ocean surface currents in several ways. The Somali Current is described as a western boundary current. It flows past the coasts of Africa and Arabia in the western Indian Ocean, near the boundary between land and ocean. Its speed has been measured as seven knots—nearly twice that of the Gulf Stream. It is an extremely shallow current but its speed allows it to move vast quantities of water. During summer monsoon months, the Somali Current moves north at its rapid rate. When the monsoons are over, the current, unlike any other yet known, reverses itself and moves southward. It is the only boundary current known that crosses the equator.

SUBMARINE CURRENTS

The existence of deep currents, oftentimes flowing in directions opposed to surface currents directly above them, has recently been established. These currents had not been known before, for there were no means available to detect them. And these currents affect the progress and stability of our nuclear submarines just as surface currents affect surface vessels. Should a submarine cruise against a strong flowing current, its rate of passage to its destination would be reduced. If, on the other hand, we can chart these secret currents and make use of them, they can speed a large undersea craft on its way as surface currents do surface ships.

Beneath the Gulf Stream on the western side of the Atlantic, oceanographers have discovered a southward flowing current that is located several hundred meters beneath the surface. And while it travels much more slowly than the northward-moving Gulf Stream, the volume of water transported southward is estimated at 500 times that of the Mississippi River. Another submarine current has been located in the Pacific at a depth of from 100 to 300 meters stretching over a distance of 3500 miles. It flows in an easterly direction—counter to the Equatorial Current—while beneath that opposing current there lies, at a depth of 1000 meters, a counter-counter current flowing to the west.

With all this new knowledge—and the suspicion that more remains hidden than is found—oceanographers are busy today revising their estimates and theories of the general circulation of the oceans. One committee has agreed that the amount of water transported by currents and mass movements of ocean water in the tropical Pacific is at least three times greater than had been believed.

These deep currents sometimes prove lethal to both

man and the life of the sea. An upwelling of cold water in the Indian Ocean near its northern reaches brings death to all life that enters it because this strange mass of water is deficient in oxygen. Just what can happen to undersea craft moving within this as yet little understood environment can well be imagined. The submarine can be cruising at a reasonable depth and suddenly stick its bow into a great flowing river of water, turning it suddenly off course. It may meet a great upwelling or sinking area of water and be thrown into a plummeting dive, just as aircraft in the atmosphere experience turbulence and loss of stability when flying through a storm area. Unlike the aircraft, though, the submarine has no warning whatever of the trouble that may lie ahead. For no sensing device yet developed can foresee the submarine's approach to such an area. And if the submarine enters an area where there is a huge, silent Niagara of sinking water, plus one or more large internal waves raging along the interface of the two different water masses, the sudden loss of stability, the turbulence and terror could take a submarine to the floor of the sea. This is exactly what happened to the U.S.S. *Thresher*.

5. LIFE IN THE SEA

PAST midnight and the sea still tossed the *Chain,* but more lightly now, on the waves that slid out of the darkness and whispered across her bow. They rolled by in endless procession. The wind had died and the clouds thinned to show a quartering moon that came and went, its light softly illuminating the jagged surface of the waters from time to time.

Lights showed on the fantail. The hydrocast was over; the winch secured. In his laboratory on the boat deck, the scientist who had gathered his data from the black depths of the water busily recorded temperatures, took water samples from the bottles and marked them for analysis the next day. He was ready to turn in.

He snapped off the lights in the lab and made his way to the officers' mess where he constructed a gargantuan sandwich of cheese and cold cuts (with pickles on the side) and drew himself a cup of hot coffee from the steaming urn.

Two sleepy scientists who had just been roused for their work straggled in. The finishing scientist joked about the dark early morning hours ahead and received nothing but noncommittal grunts from the two who were about to begin their work. They struggled over cups of coffee themselves, finished them and went on deck.

The after-fantail lights had been turned off and only the large floodlight illuminating the starboard winch and platform remained alight to obstruct the vision of the sea. A few scattered clouds floated overhead and turned to silver in the soft glow of the moon. The sound of the ship's bell came from the bridge. Four bells on the mid-watch. Two A.M.

A new winchman appeared. He yawned and took his place at the controls, checked the winch and the metering dial that showed him exactly how much cable was overside, and waited. The scientists on deck spread out the plankton net, stretching and straightening it on the mess deck below the winch. The net itself was a cone of closely and expertly woven nylon designed so that the openings between the patterns of threads were all of the same size. At the apex of the cone, they fastened a small plastic bucket in which they would catch their specimens, the myriad zooplankton of the sea.

Like the multiple forms of animal life on land and in the sea, many forms of zooplankton develop through a larval or juvenile stage and then change radically through a transformation called metamorphosis into their final and adult form. Thus a caterpillar is a larval stage while the adult moth or butterfly is the result of the metamorphosis. The project on board the *Chain* that required the taking of plankton centered on the larval stage of a particular planktonic form: *Cymatium parthenopeum.* It is a small form of sea life that develops in several places in the tropical and semitropical oceans of the world.

The reason this scientist has concentrated his attention on this small, almost microscopic bit of animal life is that if his theory is correct, he will disprove a widely held idea that such life is too fragile and short-lived to be spread all around the Atlantic basin, for example, by merely drifting here and there.

Like most oceanographers working today, this marine biologist is doing basic research. He does not know whether his work will ever have a great effect on man's practical affairs. But he does not know that it will not. On examining his project, an observer might reasonably guess that in some way, at some later time, it will indeed have some impact on men.

The current theory he is working to overturn holds

that the forms of larval and other planktonic life found on the continental shelf of Africa could not have originated in the Caribbean or along the continental shelf of the southeastern United States. In other words, animals found on the eastern and western tropical shores of the Atlantic are in separate environments that are in no way directly connected. They may have the same physical conditions for growth: temperature, depth, salinity, pressure and so forth. And this alone accounts for the similarity in the life found in these widely separated places.

This scientist believes otherwise. And he has chosen this small zooplankton form to help him prove his theory. If he is successful, the knowledge he will have discovered may assist in preserving or increasing certain forms of sea life the world around as a source of food. It may allow man to "seed" certain depleted areas with sea life. And the fact that the larva of these sea creatures can be captured and kept alive in a laboratory may allow us to grow what we want to fill depleted areas.

Over several years this scientist has gone to sea aboard oceanographic vessels and conducted plankton tows to bring from the sea the forms of zooplankton that exist off the coast of Africa, in the middle stretches of the Atlantic near the equator, off the southeast coast of the United States, in the sea near the Azores and along the Gulf Stream in the North Atlantic.

His evidence is growing that his theory and not the

prevailing one is right. He has found larva of *Cymatium parthenopeum* in many places along the gyre that encompasses both the coasts of the United States and Africa.

Wherever he has located them he has kept specimens alive in small containers aboard ship and back at his laboratory in Woods Hole for a period of as long as 14 months. Within reasonable limits this means that these larva have ample time to drift across the ocean and reach the continental shelf of Africa. Even the colder waters of the North Atlantic do not seem to harm the larva. They survive it, even though their natural environment is the warm water of tropical and subtropical seas.

With the plankton net ready now on the deck of the *Chain*, the scientist gives a wave of his hand to the winchman, who raises the net until it swings over the side. At another signal, the winchman lets out cable to a depth of about 200 meters. A signal to the bridge starts the *Chain* moving through the sea at a modest two knots. For 20 minutes the ship proceeds slowly ahead. The cable makes an angle with the surface with the weight of the net and the resistance of the water flowing through it.

At the end of that time the cable is gradually raised to the surface, still picking up plankton. Tows such as this are necessarily done during the night hours because plankton are very sensitive to light. During bright daylight hours they go to depths where the light does not penetrate. At night they rise to feed.

The net comes up through the water, becoming visible

at first as a wavering, looming shape of white beneath the black sea. When the net breaks the surface, the winchman hauls it upward quickly until the cup on the end is swinging free over the ship's rail.

The scientist easily catches the cup and draws the net inboard. He holds it a moment until his colleague lowers a plastic bucket over the side and fills it three-quarters full with a sample of sea water. Then he unscrews the cup from its fastening and plunges it upside down in the bucket of water. The clear water turns milky with the sudden presence of thousands of small zooplankton. Most are so small that anyone peering into the bucket at night on the shifting deck can see only a cloudy mass within the bucket. One or two small fish dart back and forth and several of the large plankton forms scamper around the edge of the bucket beneath the water surface.

A hose squirting sea water is turned on the meshes of the net. Its small open end is held over the bucket as the last of the gathered specimens left clinging to the cloth are washed into the bucket.

The scientist picks up the pail and walks to his laboratory, a rectangular space in the large laboratory on the mess deck. He snaps on the light, settles himself on a tall stool and adjusts a low-power microscope and several plastic dishes. He scoops up a sample from the bucket, pours a small portion into the plastic dish and peers into a completely different world.

Perhaps the first thing that strikes anyone's eye on

peering through a microscope at the live catch of zooplankton is the astonishing array of color and form displayed by the small creatures.

All the colors of the rainbow are here, flashing blues, streaks of red and violet and deep greens, yellows and oranges. These are living diamonds of the sea. They move and turn, some with jerky motions, some gliding through the water with infinite grace. Here a small shrimp contracts and expands, like an inchworm, moving through the liquid medium that is his home. There a flat almond-shaped creature rolls luxuriously, a miniature whale, flashing greens and blues as it turns. Another in a transparent barrel-shaped bit of matter moves from one end to the other of its small shelter, nervously watching for intruders or hoping to catch an even smaller creature for supper. A patch of eggs is embedded in the jelly-like barrel home surrounding it. They are light brown, small speckles of nutmeg. Some, nearly too small to catch in a glance, scoot back and forth, throwing out a brilliant pinpoint of swiftly changing colored light.

But outwardly, at least, accustomed to and unmoved by this dazzle of life, the scientist picks up a probe and a pipette. His eyes are searching for the larval form of *Cymatium parthenopeum*. When he spots one he brings the narrowed end of the pipette above it and allows water to enter, which sucks the larva into the glass tube. Then with his finger over the open upper end he lifts the tube from the slide, trapping the larva in the water withheld by the pressure. He immerses the tube in a small polyethylene container of sea water and removes his finger from the end of the tube. The water and the larva rush outward into a new "sea." The larva wiggles experimentally about in its new home—a home it may occupy for months, being carefully fed by the scientist as he determines how long it will live when taken from this point in the enormous sea.

When the zooplankton rise from the depths after dark and begin to feed, a natural question could well be: Upon what do they feed? And the answer leads to the beginnings of the cycle of life, to the small plants of the ocean or phytoplankton. In the open sea these smallest of plants are classified as yellow-green algae. Two of the smallest forms are diatoms and dinoflagellates. Diatoms are one-cell plants having various forms. They are encased in a thin shell made of a compound containing silicon. They reproduce by cell division and, like all plants, must live where there is sufficient light for the process of photosynthesis to occur. For, like plants on land, the diatoms combine, with the help of sunlight, the compounds of carbon dioxide and water to form the organic material of which their bodies are built. A by-product of the process is oxygen, which is discharged into the sea water.

Dinoflagellates are a little farther up the scale of complexity. They are also found in great variety in the sea and have two small hairlike appendages from which their names come. The flagella are not too useful to the small plants, for although they are designed to help the plant steer itself about, what feeble efforts the plant makes are lost in the sweep of wave and current in its habitat, the open sea. Like the diatom, the dinoflagellate depends on sunlight to live. Thus it too is found near the surface, or the euphotic zone, as that area of the sea penetrated by a useful amount of sunlight is called. This ordinarily is considered to be to a depth of 100 meters, and it is to this surface that light-fearing zooplankton like the *Cymatium parthenopeum* rise at night for their evening meal. Many of the zooplankton are so small that they can eat only the smaller diatoms and dinoflagellates. Others are larger and are carnivorous, feeding on the zooplankton forms around them.

The *Cymatium parthenopeum* larva that are of interest to the marine biologist aboard the *Chain* are vegetarian. In order to keep his specimens alive, the scientist has developed a "diet" of single-celled algae for his small specimens.

From these small diatoms and dinoflagellates the infinitely various life of the sea has evolved. It is as complex as life on the land, and marine biologists are not yet certain of many of its elements. The environment interchange that the scientist on the *Chain* was studying is evidence of the many gaps in exact knowledge of life in the sea.

For many years it was felt, for example, that life could

not possibly exist on the floor of the ocean basins. Men reasoned that the pressure was too great and that with the darkness and cold, fish or other forms would find the environment there too hostile.

But the argument about the pressure being too great was based on inexact knowledge. Fish near the surface or at medium depths have an oxygen sac or bladder that controls their buoyancy. Fish on the bottom have a similar sac but it is filled with liquid rather than gas. Their tissues also are liquid-filled, and thus they are not affected by the tremendous pressures encountered on the basin floor. This pressure is itself formidable and at depths of 5000 meters, for example, is over 50 tons per square foot.

Another argument against life at these depths is the complete absence of light. But the hardy dwellers of the abyss have either developed luminescent lanterns to light their way and attract other smaller fish and crustaceans to satisfy their needs for food, or have given up attempts to see altogether and rely instead on the sense of touch. They grope their way along the sea floor, many with long fins or antennae, until they encounter food. Such creatures of the deep do not rely entirely on living fish at their depth. Many forms there could not exist if it were not for the eternal falling from above of dead plants and animals. This silent, never-ending rain of organic material has been aptly termed the "snowfall" by the classic writer of the sea, Rachel Carson.

And it is a ceaseless deluge, ranging from the bodies of large fish, sharks and whales, to the lowly diatom with its nearly insoluble shell of silicate. So great has been this deluge that over the eons since the oceans formed, sediment from dead diatoms has built to great thickness. When dredging the floor of the oceans in parts of the world, oceanographers retrieve samples of what they term diatomacious ooze.

Moving up to the middle depths, the life of the sea increases. Yet even here there are relatively few forms and their number is not great. But here true fishes with oxygen bladders live within a certain range of pressure and salinity, unable to leave their environment because of their buoyancy machine which dictates the range of their swimming. Here too live giant squid with bodies up to 15 feet in length and with the two main tentacles extending as much as 50 feet into the surrounding water. Though the giant squid lives in the depths of the ocean, there have been sightings of this strange and huge animal coming to the surface to startle fishermen and give rise to legends of sea monsters.

At present our knowledge of the depths and even some of the middle levels is so small that indeed strange forms that appear can support the idea of unknown creatures in an unknown region of the earth. One fish caught in a deep net off the coast of Africa proved to be a variety of prehistoric life thought to have been extinct for over 50 million years. The only other examples of the species had been found as fossilized remains in rocks that were formed in that distant past. So the depths of the sea still contain mysteries of life waiting to be solved by the marine biologist.

Two of the sea's creatures appear to defy the strict layering of sea life according to pressure and salinity. These are whales and seals. Both these animals are mammals; whales are cetacean, the seal belongs to the suborder Pinnipedia, but both are warm-blooded like man. Both of them are capable through some mechanism not yet fully understood of diving to great depths and then rising with astonishing swiftness to the surface.

Man's experience with such sudden changes of pressure on his air-filled body has not been successful. Nitrogen from the air mixture within his body dissolves in the blood stream as he descends to great depth. If he suddenly comes to the surface, the nitrogen in the air mixture within his blood stream suddenly finds itself lodged as bubbles within his tissues. This painful situation is called the "bends" and can lead to death.

But both seals and whales, which breathe air, have some means of overcoming the problem. Whales can sound to depths of half a mile in seconds and return to the surface even faster without injury. In fact, one species of whale, the sperm whale of history in the whale-hunting days, regularly goes to great depths and finds its food in the form of the giant squid. These titans of the deep struggle far beneath the surface, and evidence of the struggle comes to us by examining the bodies of mature sperm whales killed by the whaling fleets. Across their huge brows, whalemen have seen marks from the great tentacles of the squid, scars of the underwater warfare.

Seals have regularly been taken that have had in their stomachs species of fishes of which we have no knowledge. The seals apparently dive to great depths off the edge of the continental shelf and feed on types of bottom fish that have yet to be named by marine biologists. They are referred to as "seal fish." How the air-breathing seal can adjust rapidly to the great changes of pressure remains a mystery.

Another of the as yet unsolved puzzles of sea life involves the eel, which is a true fish. These long slender animals are born in the Sargasso Sea in mid-Atlantic at between 25° and 35° North Latitude. Young eels begin their long journey to the coasts of Europe and America soon after they are born. But curiously, without training of any kind or any other mechanism of which we are certain, European eels swim unerringly for Europe, American eels for America. They never become confused. When eels reach the shorelines of the continents, the male remains in the salt water along the shore while the female swims into fresh water streams and rivers moving far away from the ocean to lakes and pools inland. For as long as 15 years, the female lives in fresh water. Then, responding to some inexplicable signal, the female begins her journey toward the ocean. When the female eels reach salt water again, the males join them and they swim to the place of their birth, the Sargasso Sea. Here mating and spawning occur and then the adult eels die. Their young, under no guidance from the past generation, set out unerringly again on one of the least understood life cycles of the ocean, toward land and fresh water once more.

Of all the sea creatures, the ones that have captured man's imagination and fancy most completely in recent years are dolphins and porpoises. These are cetaceans like the whale. They are warm-blooded, air-breathing mammals. And they have such a high level of intelligence that they astonish men who have been taught that man was far superior to other animals. Some marine biologists have maintained that it is possible that dolphins and porpoises may actually have as extensive an intelligence as men. And certainly if one judges by their calm and happy social organization, they might easily be considered more intelligent than we are. For one thing they have managed to organize themselves so that they do not wage war on one another. A sick or wounded dolphin has been observed to send out a signal and in a short time his comrades come to his aid—if necessary, supporting the crippled dolphin and hauling him to the surface for a breath of air. No dolphin ignores another in distress—a characteristic not universal in men.

If dolphins find themselves under attack, they act in concert against the common enemy. The principal method of attack is ramming the enemy with their blunt noses while traveling at high speed. And both dolphins and porpoises can travel at astonishing speeds. Some have been clocked for short distances at 30 knots. The full-grown dolphin weighs close to 1000 pounds, so when it rams an enemy, considerable injury must result. The dolphins will attack the largest sharks in the sea and have been known to attack even whales. While they may not have killed the largest whales in existence, the stunned leviathans have retreated before a concentrated attack.

In recent years the public has been amused and astonished by the antics of dolphins trained by man to obey signals and perform various acts under direction. But there is much more to it than that. For one thing, the U.S. Navy is extremely interested in discovering if we may actually be able to communicate with dolphins. The sea mammals are known to have a language for their own communication. The brain of the dolphin indicates to some researchers that he has an estimated I.Q. ranging from 100 to as high as 190. Clearly the dolphin is intelligent.

The Navy's interest in the animal is not as farfetched as it might seem. If we can establish communication with dolphins, they may well perform a great service in our researches in the ocean. We would be in the position of asking a friendly group to help us explore their natural environment. Already in training dolphins we have managed to get them to understand certain commands. They have retrieved a dummy of a man from deep in the ocean, bringing it easily to the surface. Dolphins have been harnessed literally, and one researcher believes we can train them to carry undersea measuring equipment to whatever point we wish and have the instrument activated when the dolphin arrives at the predetermined point. Other researchers see an advantage to domesticating a dolphin and then having it join a herd of wild dolphins to increase our information about the behavior of the remarkable creatures.

So far, men and dolphins have gotten along well together. In fact the dolphin shows up as being considerably more civilized than man. The animal is equipped with a fine set of sharp teeth. Without doubt it could bite an arm or a leg from anyone near it in the water. Yet in all the history of dealing with these creatures there is not one recorded instance of a dolphin harming a man. Sometimes they have had provocation when men have injured or teased them, but although the dolphin snaps a warning at his terrestrial friend and tries to avoid further injury by moving away, he has not attacked.

Dolphins are not so much docile as good-humored.

Their capacity for play seems endless and they do form true attachments to men. One case concerned a female dolphin being trained and studied in Florida. When the experiments were completed, the men released the dolphin in the warm ocean waters off the coast. A few days later, the men were cruising along the shore in a small power boat when they noticed a great thrashing about in the shallow water. They moved closer and saw an injured dolphin. They captured the animal and immediately recognized the female they had released days ago. The men took her back to the laboratory tank. On being released, she swam happily around with no sign of injury. Only then did the men realize they had been fooled by the intelligent animal. She had simply forced them to let her return where she could be with her new-found friends.

6. FOOD FROM THE SEA

As the sun was going down on the night the *Chain* encountered her storm at sea, the course of the vessel took it within sight of a group of ships seen faintly on the horizon. They appeared to be hove to. In the fading daylight, a larger vessel steamed into sight off the *Chain*'s starboard bow. She passed close to the *Chain* and was obviously headed in the direction of the more distant ships. They now appeared to be dark silhouettes against the sky. Watching from the bridge, the officer in charge could see running lights on, appearing and disappearing as the wave crests rolled between the sighting and the sighted vessels.

Through binoculars, the officer identified the large moving ship as the supply vessel of the group toward which she was steaming. All hands aboard the *Chain* had seen the ships before while traversing the Gayhead-Bermuda transect. They formed a fishing fleet, a factory ship and half a dozen trawlers. They were a small part of the worldwide fishing effort of the Soviet Union. Summer, winter, spring, and fall this group of alien vessels lies outside the waters presently claimed by the United States, just off the edge of the continental shelf, where the trawlers take fish month in and month out.

The supply ship was making a seasonal journey to bring fuel and food to the fishermen. She would stop near the factory ship, filling bunkers with fuel oil, swinging cargo nets of food from her holds to those of the factory ship. And when her holds were empty, she would begin taking aboard the frozen fish caught in the waters of the North Atlantic within 100 miles of the coastline of the United States but bound for Russian consumption.

As the *Chain* and the Russian supply ship drew closer, more and more of the ship's officers and scientists appeared on the bridge or the boat deck to get a view of her. Their interest in the passing vessel was explained when one of the officers lowered his glasses and announced that the Soviet vessel apparently was not going to conform to the international courtesy of the sea by dipping her colors as she passed a vessel of another country. A momentary irritation swept the men on the *Chain*. On a previous voyage when the Soviet ship had been sighted, the *Chain*'s pennant was run briefly down and then up in the time-honored tradition, a signal of recognition both of another nation and another group of mariners sharing common dangers of the broad face of the ocean. The Soviet vessel had ignored the signal. Now the captain of the *Chain* decided he would wait until the Soviet ship dipped her colors before extending the courtesy. He never had to send a man aft to the halyards.

Why this discourtesy? And why the recent concern about who fishes where and for how long on the open sea—a highway defended as being "owned" by every nation—as though any group of humans can own it?

For centuries men have taken their harvest of fish from the sea. Indeed some nations formed on the coasts of the continents have depended for much of their national existence on the wealth of food from the sea. In Japan, that overcrowded group of marine islands, all through history an effort has been made to gather the abundance of the sea life and process it for export as well as home consumption. With a high population density and a scarcity of tillable land, Japan has always faced the problem of feeding its citizens and struggling to amass enough business to allow it to function in the world community.

Nearer our own experience is the case of Portugal. Hundreds of years ago the Portuguese gained a reputation as fishermen. In small sailing craft, the only remaining true "men of the sea," the Portuguese set out and sailed across thousands of miles of open ocean in search of a catch. They reportedly sailed as far north as Greenland and Iceland for fish. To them the Atlantic Ocean has become a mere extension of their small country, a giant lake whose waters and vagaries they know well. On the fishing wharves of villages all around this "lake" will be found Portuguese fishermen, repairing their boats, mending nets and looking seaward trying to guess tomorrow's weather out on the sea.

In Scandinavia, settlements strung along the seacoast support large numbers of Viking descendants who contribute to the national economy by taking to the sea. Among the great whaling nations of the world, Norway, though small, has wielded great influence. The habits of the whale have been studied by Norwegians for centuries, the vulnerable places on each species for a quick kill and the best approach to leviathan in the waters of the antarctic—all this lore has been passed down from generation to generation. Today Norwegian whale hunters roam the Southern Ocean circling the Antarctic Continent. The small powerful ships of the whaling fleets, known as catchers, are most likely to have "Norski" on the bow manning the modern gun that fires the harpoon into the whale.

To a lesser extent, other nations with a border defined by the encircling sea have been and are today dependent on the fish of the ocean for much of their food and income. All the coastal nations of Europe and South America have growing fishing industries. In more primitive areas, local fleets support entire villages of emergent populations, and men and women and children live by the rhythm and the mood of the sea.

The coastal regions of the United States follow the same pattern. No small percentage of the families there gain their livelihood from the sea. Tuna fleets put out from West Coast ports. Shrimpers leave the wharfside from Gulf ports, and lobster boats make the rounds of the buoyed traps in the cold waters off the Maine shore.

Scientists today are uncertain just how large a crop of food can be harvested from the sea without depleting the stocks of the great oceanic storehouse. At first glance this storehouse would seem to be inexhaustible. Estimates of the total annual production of sea life, from the lowly phytoplankton to the largest sea creatures, range up to 800 trillion pounds. And today, United States fishermen take only about seven billion pounds of food from the oceans each year. Worldwide fishing, which has doubled in the last 13 years, takes only about 120 billion pounds per year. Yet there have been examples of overfishing in certain rich areas and until further research locates and tracks the life of the sea in great detail, such overfishing will continue. Modern fishing methods are proving to be much more efficient than the traditional ones, and while they assure good catches, they can help to produce local depletion.

Over the centuries that men have fished the sea, they have developed methods of gathering as many fish per trip as their ingenuity allows. The principal methods in common use today are: trawling, seining and drifting.

The trawl is an open-mouth net bag that is dragged along the sea floor for bottom-living fish. The trawl is lowered to depths of up to 1000 feet, and flounder, plaice and skate are taken. A seine is also an open-mouth net bag but it is lowered to middle depths and pulled behind the fishing vessel—hopefully, through a school of fish. The drift method employs a long net with buoys along one edge and weights along the other. It is let out as the fishing vessel makes a large circular turn in the water. The net hangs down like a huge curtain to entrap fish, like the herring, that rise to feed just beneath the surface.

In days of sail, the progress and maneuverability of fishing vessels was slow. But with the invention of powered vessels, more nets, quicker trips and larger catches resulted. And yet for decades the system and the habits of fishermen remained the same. In each locality, knowledgeable old fishermen seemed to know where the schools of herring ran, where the best and most abundant cod or haddock could be taken. So tradition was formed and men followed it, going each season to special fishing grounds and trusting to luck to bring back a large catch.

Then World War II spurred research in sea measuring and communication equipment. Sonar, underwater sound impulses that act like radar in the atmosphere became a means of defense against submarine attack, a method of learning where the enemy was located. But frequently things went wrong with that primitive gear. Odd reflections came back to the watching officers and at last it was realized that the "enemy" frequently spotted was a school of fish.

Not long after the end of the war, commercial fishermen began locating fish by two methods, new at that time: sonar and airplane reconnaisance. The herring boats that put out from ports in Maine were directed by light planes that flew low over the coastal waters looking for the fish. As soon as herring were spotted, radio messages brought the boats to the scene and drift nets were put out in the path of the school.

Expensive sonar sets were installed on modern fishing vessels and the fish were tracked as they never had been before.

Yet even with these modern tracking devices, the majority of fishing fleets from most countries continued to use the traditional methods. Habit, long ingrained, sent the small ships to the same fishing grounds even though they might have been overfished the season before.

Some nations could not afford to rely on habit, however, and they began to draw on the knowledge of their oceanographers and to launch new studies to help them find fish. A good example is the island nation of Japan. A hint of the problem the Japanese were to face came to them over 35 years ago when a cold current known as the Kuroshio Current was pushed out to sea farther from its usual path along the coastline. The cause of this shift of the current apparently was the creation of a small inshore gyral or eddy. For many years the Japanese fishing fleet had gone to the area formerly occupied by the current and found good fishing there. Then one year they returned without a catch. Year after year, the hopeful fishermen returned to the same spot out of habit, only to come back empty handed. No one analyzed the current shift and no one knew that the fish they sought required the cold water of that current to survive. And since three-quarters of the protein needed by the Japanese people comes from sea products, the barren years struck a heavy blow at their diet.

Then the Japanese entered World War II and by the end of it had lost all their shipping, including all or nearly all of their fishing fleet. When Japan surrendered, her leaders realized they would have to take drastic measures if they were to feed the nation. It was a struggle for survival and the Japanese did survive.

They studied the problem of their fisheries. While the oceanographers contributed their knowledge, surveys were made of currents, temperatures, salinities and densities in all the waters within range of the islands of Japan. While these studies went forward, ships were being built. Thirteen years after Japan's surrender, her fishing fleets took two billion pounds more fish from the sea than they ever had before.

The most modern approach to fishing in the world today has been applied by the Soviets. Billions of dollars spent over the last 15 years have produced the most modern fishing fleet in the world. The oceans of the entire globe are open to them and they are sending fleets to all of them. The usual fleet is one much like that sighted by the *Chain,* a headquarters ship, complete with sonar, loran, radar, supplies for an extended cruise. It is, in fact, a miniature world at sea surrounded by the small trawlers that put out for the actual catching. When the smaller ships are loaded, they come in and discharge the fish directly into a freeze plant aboard the headquarters vessel. Then they put off once more to catch more fish.

Their modern methods were nowhere more in evident contrast with outmoded methods than several years ago when American fishermen from the northeast United States were going to their usual fishing grounds off Newfoundland. American ships—small wooden vessels with little more than a ship-to-shore radio in the way of modern instrumentation—reported small catches. They could not find the fish. Luck was not with them. The Soviets did not rely on luck. Their up-to-date sonar located the schools and caught up with them. The Soviet's trawlers brought in tons of fish while the Americans complained of small catches.

Modern studies, largely conducted by other countries, indicate that when fish are no longer found in traditional locations they are not necessarily all gone. They simply have moved to a new location. For the schools that swarm the ocean are sensitive to slight changes in water temperature, to currents and to salinity and planktonic content of the water. They also learn to avoid areas where their numbers are reduced by overfishing. This instinct to survive has not been taken into account before. But all species of animal life have this instinct whatever their intelligence level. Otherwise they simply would not have survived to the present time. They adjust to change, and if the change they encounter is the presence of many large nets that deplete their numbers considerably, they do not return to that spot.

The Soviets discovered this and stopped depending on traditional fishing grounds to find fish. They now hunt them with every modern sensing device that technology offers. Britain and Norway, Sweden and Japan are all major users of modern technological advances and oceanographic data. The United States has a long way to go to catch up with the major fishing powers of the world.

The worldwide race to catch fish has raised many new political problems that have yet to be solved. The coming of these problems was indicated when tuna boats of the West Coast, fishing in traditional open-sea waters off Central America, were fired upon by gunboats of a Central American republic. The question of sovereignty does not exist in most cases in the open ocean beyond the edge of the continental shelves. But within that region, controversy continues. Here much of the life of the sea exists and here is where the questions of territorial right come up. Some countries claim that part of the ocean as their own that lies three miles from shore. Others claim six and still others twelve. Some legal experts have suggested the edge of the continental shelf as basis for a claim of sovereignty. And a number of nations are now claiming 30 miles of ocean surrounding their coasts. Since the continental shelf is not precisely defined, however, and since it also varies considerably in width from place to place around the land masses of the world, there is plenty of room for disagreement.

The struggle for a more efficient means of extracting life from the sea is not entirely the result of private greed. Another factor has long been at work and is now, in some parts of the world, reaching undreamed of proportions. Indeed many statesmen feel it is the primary problem facing all human life on the planet today: the population explosion.

World population has increased sharply and still is going upward. Tillable land, the means of raising food on it and the means of distributing it are simply not able to keep pace with the cry for food from mankind. Drought and crop failures have intensified the trouble.

It is this problem that has brought many scientists to work in the field of oceanography with specific interest in extracting food from the seas. They are seeking not only animal life but plant life as well.

Experiments now in progress are attempting to turn some of the phytoplankton and kelp of the sea into nutritious food. But now the problem is quantity, and the green algae in the sea is being tested and mixed with other products to produce a baking flour that will be edible, palatable and nutritious. It is estimated by one such researcher that only five million acres of algae cultures could satisfy the protein needs of the population of the United States. At the moment we need 300 million acres of land to produce an equal amount of protein food for our diet. Already a flour has been produced from green algae that can be made into wafers. If we can manage to transform the uncountable billions of some forms of planktonic life into food, we will have gone a long way toward solving the nutrition problems of an exploding humanity.

But these experiments do not provide food today for the hungry millions. Modern methods of catching fish may be able to do the job more rapidly One up-to-date method that used an electric current of 20,000 amperes flowing from a submerged electrode was developed in

Germany. For reasons not yet fully understood this flowing current attracted fish within a hundred feet of the electrode. When the fish gathered closer, they were pumped up a huge suction pipe, water and fish together—thousands and thousands of fish. The fish passed through the pipe, were killed instantly by an electrocuting mechanism and deposited in freezers in a matter of seconds.

The Soviets have fished the same way, using light and sound to attract fish to the submerged end of the large suction pipe. Some experiments even have used music as a lure. It worked astonishingly well. The fish flocked to the source of the sound and were observed swimming round and round it—moving with the rhythm of the music. They seemed to have been hypnotized.

These methods of hunting in the seas could be helped greatly by knowing where, in the millions of square miles of ocean, fish were located, where they were headed and when they might be expected to arrive at a particular fishing ground. Attempts by marine biologists are now under way to gather such data, but it will take many years.

The process is one already familiar to ornithologists: tagging. As birds are banded and released to follow and chart their migration patterns, so fish are now being caught and marked or tagged and then released. Days, weeks or months later when a fisherman manages to catch one of the tagged fish, a report to the Oceanographic Institute that tagged it gives one more small but important piece of information to chart the fish's journey. In one recent year, for example, marine biologists at Woods Hole tagged 4000 specimens and released them. The waiting biologists received 213 "replies," which was the best return they had ever had. Japanese fishing experts are doing the same thing, as are scientists of other nations. Eventually there will be enough data to allow computation of their meaning and the revelation of patterns of fish movement that are still largely unknown except in a few isolated cases.

It has occurred to many scientists and fishing experts that it might be profitable to farm the sea just as we farm land. In the case of kelp this has already been done in shallow coastal waters. The crop is harvested by mowing under water when the kelp has matured. Fish have been farmed to some extent in the Philippine Islands where experiments of a crude sort have produced good results. Fishermen actually feed the fish by sprinkling areas of the surface with flour or ground corn. Gradually the fish realize that extra food awaits them in the seeded spot and they are easily netted from time to time.

There are steps beyond this. Building actual underwater fences is not practical from an economic standpoint but transmitters can be set beneath the surface and with directional sound waves, a curtain or fence of sound could enclose the fish and keep them as safe as in a weir until the time came for harvest. This way, areas of bays or shallow seas could be stocked. The fish could mature in their fenced pasture and then they could be lured to a great suction pipe by an attractive rather than a repelling sound.

As in all experiments with natural environment, a serious danger exists in an unconscious or uncontrollable disruption of the total environment of the fish, their ecology. We know little of this side of the problem and by acting in one direction we may disrupt the ecology and defeat ourselves in the attempt to control sea life and make it subject to our needs or desires. One curious example came to light in the recent dissension among Soviet and American fishermen. While the Americans were searching for cod and haddock deep under water, the Soviets were taking herring near the surface. As herring disappeared from the area after intensive fishing, the Americans found their catch of cod and haddock was sharply reduced.

The disruption of environment by man is not a new danger. It happens over and over when man focuses his attention on one specific problem and ignores all the surrounding known and possible circumstances. Our convenience or our greed has often violently upset natural cycles, and we have always suffered in the long run. Controversy rages today about the widespread use of insecticides, and not until the late Rachel Carson's book *Silent*

Spring drew attention to the problem did any organization begin to act. Today doctors throughout the world are worried about the effects of antibiotics. The wonder drugs have saved millions of lives in the past. But their widespread use is producing a medicine-resistant form of life that carries disease. The natural creatures were at first killed by the antibiotics but have now developed immunity against former killers and in fact can "immunize" fellow bacteria, making them safe from attack by several of our most powerful wonder drugs.

Can we upset the ecology of certain species in the sea? Without any doubt we can, in many ways, and the easiest to grasp and witness is the actual killing off of certain species of life that have attracted man for food or to serve him in some other way. One of the most dramatic of these catastrophes is the near extinction of the fur seal found in Alaskan waters. When the United States purchased Alaska in 1867 there were an estimated 4,500,000 seals living in the Bering Sea, principally on the Pribilof Islands—which have long been called the Seal Islands. By 1912 seals had been killed to such an extent that the most optimistic count put their number at a little over 200,000. These animals were not used as food but were prized for their fur, which was made into women's coats. By the time the seal herd had diminished in 1912, men began to worry about their extinction, and international regulations were agreed upon so that the species would survive. It has done that, but it is a luckier species than many others that men have obliterated.

Greed alone is accounting for the obliteration right now of one of the most extraordinary species of animal life the world has ever known: the whale. The process of obliteration is not seen by the average citizen, and little publicity accompanies the slaughter, so it is more than possible that within a few years these largest of the world's mammals will be gone.

The beginning of the obliteration of this species lies back in the great whaling days of the early and middle nineteenth century. Then, sailing ships from all the major nations of the world hunted the whale. In the United States the ports of Salem and Stonington, New Bedford, Edgartown, Boston and New Haven, plus many others, became world famous. Ships from those ports roamed the oceans of the entire world in search of whale. The value then of the whale was in its oil, for this was before the discovery of petroleum, before the advent of kerosene for lamps. Whale-oil lamps were common, and to light the world, men hunted whales. There were fortunes to be made in one voyage lasting for as long as three or four years. Barrels of oil came back in the ships but only after the whalemen battled leviathan in his own environment and frequently at the risk of their necks.

The small boats and crude methods of catching whale in those days confined the efforts of the hunters to those species of the animal that did not swim too fast to be overtaken, harpooned and then lanced to death, and would have the courtesy to float after being killed. Thus the bowhead whale and the right whale were the prime targets of these early hunters. The sperm whale was larger and was hunted also, but this species was more dangerous. It had been known to turn on its pursuers and attack not only the small boats towing after it, but also the whaleship itself. When Herman Melville recounted such an attack in his great story of whaling, *Moby Dick,* some readers, unacquainted with the sperm whale and its size and temper, disbelieved the author. But shortly after *Moby Dick* was published in 1851, a report came to New England ports that a ship had been sunk in just such a manner.

The sperm whale is carnivorous and aggressive. Its natural prey is the giant squid and the octopus. It has teeth and can use them. The sperm whale belongs to a suborder of the species generally known as toothed whales. The bowhead and right whales are what are called baleen whales. These whales do not have teeth in the usual sense but rather have a fine mesh of bony material (baleen) in their mouths through which they strain ton upon ton of plankton from the water for food. These largest of the ocean's mammals feed on the smallest forms of sea life. They spend a great deal of time straining the plankton-rich sea water of colder latitudes through their mouths just in order to stay alive. Yet so vigorously were the right and bowhead whales hunted and killed that they nearly became extinct. They are supposed to be protected today by international agreement among whaling nations but reports have come of illegal killing of several of these whales each year. So rare are they even now, after years of protection, that whalers sighting the right whale report it as an unknown species.

During Melville's time when men rowed whaleboats in chase of the animals, the largest of the whales, the blue whale, was rarely attacked. It was too large, too fast and too powerful. And if by mistake a harpooner put his iron in a blue whale (which was also then known as a sulfur-bottom whale) the line and frequently the boat were demolished in short order. The ancient cry of "a dead whale or a stove boat" was most often the latter when blues were hooked onto. But with the advent of steam, of explosive harpoon charges and of modern factory ships, the blue whale became the prize catch of the whaling industry. Measuring on an average 85 feet in length and yielding 40,000 pounds of oil, the blue whale actually sets the standard for other whales. Thus a whale like the finback that yields only 20,000 pounds of oil on the average is considered to be only half of a "blue whale unit." And the season's catch is counted in blue whale units.

Since the beginning of the twentieth century, an estimated 325,000 blue whales have been caught in antarctic waters. Each year the number caught are fewer and fewer. The average size caught is also smaller and there is great concern now that the species has been killed to such an extent that it cannot recover. Even if the whaling nations agreed to stop killing blue whales entirely, there is only a slim chance that this species will survive. For when a species is reduced in number below a critical point, it cannot breed enough new stock to replace members that die from natural causes in the sea. When that occurs, the species is doomed.

Conservationists have estimated that with a healthy species of sufficient number, 10% to 15% of that species can be hunted and killed without endangering its existence. But in the 1950's, over the objections of just such a conservationist committee, the whaling nations, then principally the Soviet Union, Japan, Norway and Great Britain, went ahead and killed a greater number than the conservationists advised. The result was that the catch of blue whales was reduced and the whalers turned to the killing of other species, principally the sei whales, minke whales and the sperm whales. They were working for economic profit and it did not matter to them that they were killing whales in such numbers that they were endangering the entire species.

Then in 1960 the situation became so acute that the International Whaling Commission decided to create a committee of non-whalers to determine the whale populations of the various commercially hunted species and to ask for advice on the quotas to be set for the coming seasons. The committee was made up of population experts from three countries, and their report was due in three years. During the time the committee was studying the problem, the catch in antarctic waters dwindled. Blue whales caught shrank from 1740 in the 1960–61 season to under 1000 in the 1962–63 season. To make up for this the whalers killed off finbacks and sei whales to fill their quotas for the owners, who were thinking solely in terms of immediate profit.

The committee warned the whaling industry that catches would continue to dwindle, but some whalers and whaleship owners simply did not believe that the three-man committee knew what it was talking about. The 1963–64 season, the committee claimed, would see the taking of a total of only 8500 blue whale units (compared to 16,000 blue whale units in the 1960–61 season). They also predicted that to take even the 8500 units, whalers would have to kill 14,000 finback whales and a negligible number of actual blue whales. When that disastrous season was over, the entire industry, composed of 16 factory ships with their attendant catchers, had taken 8429 blue whale units and had killed in the process 13,870 finback whales.

There was little question but that the committee knew what it was doing. Its members strongly recommended that the whaling industry limit itself in the 1964–65 season to 4000 blue whale units. They further recommended that in 1965–66 the quota be 3000 units; for the 1966–67 season, 2000 units. This reduced number of kills in those waters was necessary, said the committee, just to keep the whale population at its present level.

Since the late 1950's, Great Britain has dropped from the scene as a pelagic whaling nation. She no longer sends a fleet to the open ocean to hunt whales. But Japan, Norway, the U.S.S.R. and The Netherlands still do. These nations rejected the committee's recommendation and agreed among themselves to take no more than 8000 units —twice that recommended for the 1964 season. But when the season ended they had not reached their quota, and more impressive, only 20 blue whales had been caught by the combined fleets. Finback whales also were scarce, and the fleets killed more and more sei whales in a vain attempt to get 8000 units.

On their return from antarctic waters, the whalers learned to their dismay that more whales were taken in the North Pacific than they had managed to get in that once prime whaling area, the Southern Ocean surrounding the Antarctic Continent.

Only one factor has forced the whalers to reconsider their position in the whole matter of whaling: economics. When it became unprofitable to kill whales, they slowed down or stopped. But by this time the entire question of whether or not whales could continue to exist on the planet had been raised. The blue whale, the true leviathan of the deep, may already be doomed. The finback may also be on the road to extinction. For the present stocks of these whales in the oceans of the world may no longer be sufficient to insure survival of the species even without hunting by men. The sperm whale is still hunted, and in the last dozen years over 250,000 of these great mammals have been killed.

One expert estimate is that even if strict whaling quotas are applied and the whaling nations stick to them, nearly a century will have to pass before the herds of whales are rebuilt to the size they were only 15 years ago.

Can man affect the existence of an entire species? Indeed so—even among the greatest animals ever to appear on the earth.

Yet to many people who see the wealth in the ocean, it remains inconceivable that the supply of a plentiful species will ever be fished out. What they do not stop to consider is the enormous number of animals involved in the cycle of life in the sea. They see only the great schools of herring or tuna or the vast millions of square miles of ocean surface teeming with planktonic life. What they do not see is the mortality rate in the production of just one fish.

Scientists have estimated that it requires 10,000 pounds of diatoms and dinoflagellates (phytoplankton) to produce 1000 pounds of zooplankton. The 1000 pounds of zooplankton can be transformed into 100 pounds of herring. These herring can raise 10 pounds of baby tuna which finally becomes one pound of adult tuna available for men to eat. The ratio here is 10,000 to 1, so when men gaze on the life in the sea and imagine that the supply of food there is unlimited, they should keep in mind that man has destroyed wherever he has turned his attention and greed—sometimes beyond repair.

7. THE WEALTH OF THE SEA

THE ship is an atom, a dot, a point in the closed envelope of the black darkness of night, rising and falling on the now dying swells of the sea. It wallows in insignificance and is something only to the men who are awake.

There are few of them. On the bridge the watch officer leans against the rail, looking aft into the cone of glaring light. In the laboratory the marine biologist has finished gathering larva for his collection. He takes the half-filled specimen bucket, milky with thousands of plankton, and strains them out with a sieve. He dumps the sample of life in a preserving solution and covers the container. Then he snaps off the microscope light, covers the instrument with a plastic hood and goes on deck. With a line attached to his specimen bucket, he tosses the pink pail into the ocean and rinses it. He walks across the wet, gray deck plates and turns on the hose to rinse the plankton net that lies near the rail. When he is finished, he hangs the net up to dry in the cool night air. In the circle of light shining on the water, two small squid rise like infant ghosts from the depths, attracted by the commotion of the splashing bucket. They lurk four or five feet beneath the surface and then, sighting or otherwise sensing some unseen prey, flash to the surface, their two longest tentacles reaching ahead of them, grasping the food (the plankton) breaking the surface of the water momentarily. Then they sink down almost below the reach of the light, to wait again.

The biologist yawns and leans over the rail a moment before turning in, watching the little squid. They come up once more and seize food and go down ready to repeat the process, watchful and wary, swift animals from the other world, the ocean.

While the scientist watches the squid, the edge of the circle of light on the water is blurred. The brilliance of the light seems to fail. The biologist looks up in surprise and stares toward the horizon where a strange diffused light makes the horizon visible, a lustrous, wavering leaden line, dark against the lightening sky. Dawn comes swiftly at sea.

* * *

The wind had died and the sea was silent except for a whisper now and then as the crest of a long swell broke against the hull of the *Chain*. The air was cool with a salty morning freshness and the clouds that had made the night so black were breaking. The first light was a dim gray, but as the sun came nearer the horizon, faint pink and pearly hues touched the sky.

The biologist walked slowly forward to the bridge and notified the watch officer that it was time for the deep dredge. Then he went below and wakened the scientists working on that particular project. Then, without another word, he went to his bunk and turned in, falling instantly asleep.

The first thing that greeted the new team of scientists when they emerged was the clatter of pans in the galley, where the cook fought his way toward the preparation of breakfast. The mess boy walked dazedly around the officers' mess, setting out plates and silver.

Pausing only for a quick cup of coffee, the scientists moved through the messroom and the laboratory aft to the open deck. There they were greeted by the young representative of the marine instrument company, who was already checking the spring-driven "trap" door that was under test on the dredge. When he had completed his inspection and pronounced the mechanism in perfect order, the men stooped and lifted the heavy metal dredge, carrying it with many grunts and exclamations to the after edge of the fantail, where they lowered it to the deck. One man fastened the cable end to the dredge. A crew member cleared the cable and nodded to the new and as yet only half awake winchman. The winch motor sprang to life and the rumble of the great drum was, for a moment, the only sound that broke the silence. The cable tightened and the end of the dredge was drawn slowly upward until a hand signal from the crewman stopped the winch. The dredge hung vertically, only its after-end resting on the deck. Several pairs of hands steadied the dredge as the ship rolled slightly in the swells. It was an ideal day for the lowering of a dredge in the open sea to a depth of 5000 meters.

Cautiously, the men inched the end of the dredge to the outermost lip of the stern and gingerly pushed one corner of it over, then the other. The crewman waved his hand again and the drum rumbled. Still maintaining contact with the upper end of the dredge, the men guided it over the stern. When it cleared the ship and sank slowly into the water, the men watched it until it disappeared on its next journey to the bottom of the sea.

For a moment there was no sound but the winch drum letting out cable. The men stared at the spot where the cable entered the water as though by mere mental concentration they could follow the unknowable path of their instrument to its hoped-for location on the bottom of the sea.

The cable was payed out slowly, at the rate of 50 meters a minute. To reach a 5000-meter depth and allow enough slack to drag the dredge across the bottom, the scientists had learned they had to release 7500 meters of cable. It would, therefore, take two hours and thirty minutes for the cable to run out to the desired length. The scientist in charge waved his hand to the officer on the bridge who was watching the proceedings intently. It was the signal for a very slow speed ahead—two knots—so the cable would not foul on the sea floor when the dredge first struck bottom. Fully awake now, the scientists returned to the mess for breakfast.

While they ate they discussed the occasions when, after all preparation and precaution, they had finally brought the dredge back on deck with absolutely nothing captured in the nylon mesh net within the metal frame. On those disappointing occasions there were arguments over the cause of failure, speculations on what actually had happened in the black depths of the ocean into which they could not peer to learn directly how the dredge was working.

The purpose of the dredging project was to capture the bottom life in the sea by skimming it from the top of the sea floor and sediment. Since there might well be some life existing within an inch or two of the floor surface, the dredge opening had been designed with heavy steel lips, or flanges, that could be adjusted to "bite" more or less deeply into the bottom. The question of what type of bottom they might encounter in any given place could not be answered directly. Also, since the scientists intended to have the ship tow the dredge for as much as half an hour at two knots, they might easily pass over several different kinds of bottom formation: diatomic ooze, sand or rock, or any combination of these. The dredge might fall into a canyon in the sea floor. It might encounter a sudden cliff-like rise along its path and be wedged among a jungle of rocks. There simply was no way of knowing.

It is true that the general character of the ocean bottom in the part of the Atlantic in which they were working is well known. It is a flat plain for the most part, but even flat plains have irregularities, rocks, indentations and unknown areas, in any one of which the scientist could lose the dredge and all specimens it might have gathered.

There is no greater disappointment among scientists than an experiment or a data-gathering attempt that fails. If the cause of failure can be determined, the chagrin is short-lived and steps may be taken to correct the faulty equipment or method of employing it. But when no reasons can be found for the failure, disappointment is acute and gloom pervades the quiet comments among the men. So, on any lowering of the dredge, the unspoken thought in every scientist's mind is, "Will we get anything this time?"

With the slow paying out of cable, the slow 30-minute dragging of it along the sea bottom, and the equally slow recovery of the instrument, there is more than enough time for gloomy thoughts to enter and linger in their minds.

When the dredge cable is fully payed out to the extent of 7500 meters, the ship is steered along a 5000-meter contour over the sea floor. The contour is followed by paying close attention to the fathometer on the bridge which reports instant and accurate readings of the water depth.

After the 30 minutes has passed, the ship heaves to and the winch drum again rumbles. The wet cable moves slowly up over the stern and passes over a large steel single-sheave pulley fastened to the peak of the after A-frame.

The scientists check the winch-drum meter. When the dredge is within 500 meters of the surface, the nervousness of the men increases noticeably. Remarks are shorter, more telegraphic, interest and anticipation infects the crew. The crewman who has the responsibility of guiding the dredge safely to clear the stern is leaning over that endmost part of the ship hoping to be the first to catch a glimpse of the metal cage. Several of the scientists join him until nearly the entire group is perched precariously on the edge of the fantail, straining for a glimpse of the instrument that has been to the depths of the ocean.

At last there is a cry that comes simultaneously from several of them, "There it is!" and a moment of tense silence before it comes close enough to the surface to report a catch or another possible failure.

"The door's shut," someone shouts, and then, in this case and on this occasion, it became evident to all who were watching, that the dredge had, indeed, "got something." The sides of the cloth net bulged. When the dredge cleared the surface and for a moment hung free above the stern for all to see, everyone realized it had been crossing a muddy part of the ocean floor. Mud coated the dredge and streamed in the escaping water down again into the sea. It was a dull, colorless mud and when the dredge was tugged and hauled inboard so it could be lowered to the deck, the scientists knew that while they had certainly netted a great deal of real estate, it was not what they had had in mind.

To overcome failure to make a successful haul, the scientist had adjusted the lips of the dredge in the extreme open position. This was done to insure a maximum "bite" into the sea bottom and, in this case, the bite had been too extreme. The dredge was filled with one of the sea's most valuable commodities: manganese nodules. They were round rocks the size of tennis balls that covered the part of the sea floor over which the *Chain* had hauled the dredge. Of great interest to industry and geologists, they had no value for the marine biologists studying the life of the sea floor.

When the dredge was finally emptied, the nodules filled four 50-gallon tins. The mud was carefully screened for any signs of microscopic life and the resulting wash stored in the familiar plastic containers.

* * *

The manganese nodules brought to the surface that day provided only a hint of the possible wealth of minerals that men have begun to notice and search for on the floor of the sea. Scientists now believe that the ocean bottom is covered with vast areas of these and other valuable minerals. The manganese nodules themselves contain other elements, but the per cent of manganese is of special interest. It varies from 20 to 45 per cent, depending on the area from which it is taken. The nodules also contain cobalt and iron in interesting quantities. Investigation is still going on to determine the areas of richest deposits. Industrial companies, such as Reynolds Metals, are exploring the coastal waters of the United States in deep diving submarines in search of these rare metals.

Such search has paid off. Working along the sea floor off the coast of South Carolina, the *Aluminaut,* an experimental undersea research vessel being tested by Reynolds

Metals, has discovered much more than a sea floor covered with manganese nodules. It has come across an entire ledge of the material that the searchers described as a highway. Evidently, the Gulf Stream in that area keeps sediment from collecting on this vast, rich sheet of manganese. The *Aluminaut* was actually equipped with wheels that allowed it to ride along the "paved" sea bottom, "as though," one researcher reported, "it was on a country road." With the aid of mechanical arms fixed to the undersea craft, a 198-pound slab of the "road" was retrieved and carried to the surface.

Company geologists believe that the sample found off the southern United States will easily exceed the minimum commercial standards for manganese ore. The element is rarely used in its pure state, but each year the steel industry buys over 1,700,000 tons of manganese to strengthen steel. It is also used in the manufacture of fertilizers, in photographic materials and in the chemical and pulp and paper industries.

For years the presence of such nodules as the *Chain* brought up was known by scientists. However, their interest in the ore was centered around answering the question: How does the nodule form? Only recently has a scientist answered that question.

Examination of the nodules shows that they are begun and grow through the action of a small marine organism. Upon breaking nodules apart and examining them carefully, the scientist found in the center of each nodule a tiny one-celled animal. To its shell was attached an even smaller organism that fed on the animal. It was this small parasitic creature that drew the manganese from the sea water, extracting it and forming a thin coat of the material around it. The coat continued to build until a large nodule formed. No one knows yet exactly how long it takes to form a nodule, but the comforting thought is that the process is going on all the time. So there are production plants in the sea that can supply men for years to come with manganese and other minerals.

The prospect of minerals from the sea has caused a great deal of excitement recently among both scientists and industrial concerns. Quick calculation shows them that each cubic mile of sea water (of which there are an estimated 350,000,000) contains over 160,000,000 tons of dissolved or suspended material. Much of this is made up of valuable minerals.

Some areas of the oceans are much richer than others in these dissolved minerals. We already know that certain waters in the Red Sea, for example, contain fantastically high percentages of minerals. Some Red Sea water has as much as 50,000 times the average percentage of such elements as iron, copper, lead and manganese. The Mediterranean is another body of water that has higher than average percentages of minerals.

At present it is not economical to "mine" the sea for

all of the useful elements, but already we are extracting sodium and chlorine from dissolved salt, plus bromine and magnesium. According to one oceanographer's careful estimate, approximately 3.5% of all the sea water in the world is dissolved elemental matter. Platinum and gold, vanadium and cobalt all may soon come from sea water itself.

One element, uranium, which is of great interest to all nations, is also present in sea water. Reports have come from Great Britain that an organization there, the National Physical Laboratory, has developed a method of extracting uranium from sea water at a cost that compares favorably with that of extracting it from low-grade ores. High-grade ores still draw the attention of nations needing uranium, but the time probably will come when they will turn their attention to the sea as a source for this valuable element. Nations that have no high-grade ore may easily begin now to take it from the sea.

Surveys of the wealth to be found in both sea water and the ocean floor itself have already produced several new industries that do nothing but focus on the business of mining the sea. A spokesman for one of them, Ocean Resources, Inc. in California, has indicated how industry is approaching the problem. Mineral deposits in the sea, he said, are looked upon as occupying several regions—beaches, sea water, the continental shelf, the deep sea floor, and rocks underlying the floor. Continental shelves, for example, yield "calcareous deposits, glauconite, placer deposits, sand, gravel, and phosphorite." He cited a new type of sea-floor phosphorite deposit discovered off the coast of Mexico. In a miner's phrase, the only "overburden" on this deposit is the water of the sea itself. The accidental discovery of the phosphorite revealed that in this one site alone 200 billion tons of phosphate materials —enough to supply the world for 4000 years at the present rate of use—lies waiting for exploitation.

One mining operation now going on commercially is the retrieval of diamonds from the sea floor. Off the coast of Africa, a ship is lowering a large hose to the ocean bottom and sucking up bottom sediments which are washed and searched for diamonds. The results are better for each dollar spent than those of the traditional diamond mines on the African Continent. Less spectacular, but equally important, is a new sulfur mining process being carried on in the Gulf of Mexico. From a steel platform, heated sea water is pumped to a deposit of sulfur found below the sea floor where it melts this valuable element. The liquid solution is then pumped several miles through a heated pipe to a shore plant where the sulfur is extracted and shipped. On the continental shelves around the United States many rich deposits of other valuable elements and compounds have been found. These shelves are nothing more than drowned coastal areas of the continent and would have the same character as the land itself. There is no reason to doubt that this vast new area just along our own continent contains resources that can be useful to man.

One of the earliest of these materials to be exploited is petroleum. The first search for it began nearly 20 years ago in the waters off the coast of Louisiana. At the outset it seemed logical that oil might be found beneath the offshore waters since the existing oil fields reached to the shore. Drilling rigs were towed out into shallow water and sunk in place to be used as platforms to hold the drill and tower. From this small beginning the oil industry pioneered in the first important development of extracting wealth from the sea floor.

Today geologists believe there are hundreds of billions of barrels of oil beneath the seas to be tapped between the shoreline and depths of 300 meters. Originally starting in the shallowest water, oil drillers have moved out now to greater depths. Equipment presently at work includes over 180 drilling rigs working around the world on the continental shelves of 60 countries. There are essentially but two kinds of rigs—fixed and floating. From the crude sunken barges of several years ago to the present fixed rig is a long step. One rig today stands on legs 350 feet tall and supports a drilling platform that is half an acre in extent. It is drilling in water that is 170 feet deep. All such rigs must be capable of withstanding storm waves and wind. They have a disadvantage in that their construction and removal for transfer to another location is expensive and difficult.

By far the most ingenious rigs are the "floaters," as they are called. The earliest of these was nothing more than a vessel fitted with a drill and anchored in place in shallow water. But as depths increased, anchoring the vessel became more and more difficult. The first obvious solution was to design a vessel that could be kept in place by means of angled propellers, both fore and aft, to react to wind and tide effects, providing a counterthrust that would keep the vessel in position over the drilling site. Such vessels have been constructed and they work well, but they cannot eliminate the pitch and roll of the ship in stormy weather. To overcome this difficulty, the oil companies have devised a semisubmersible vessel. This is actually a floating ship, but it can be partially flooded to send it down beneath the surface and far enough under the waves so they have little or no effect on the vessel. Such semisubmersibles have operated safely with a sea of 50-foot hurricane-inspired waves rolling above them. So firmly do the oil companies believe in the search for oil under the sea that they are spending between five and ten million dollars for the construction of a single such vessel.

Eventually the search for oil will force the development of equipment that will rest on the sea floor itself. Fed by air from the surface, fixed to the sea bottom and housing the entire rig and the men necessary to work it, the sunken unit will be an undersea community. Pipelines will carry the petroleum to the surface for loading aboard tankers, or it will be sent directly ashore through various pumping stations across the sea bottom.

* * *

Not only is the sea capable of supplying profit to men who dare to seek its wealth, it also can give life in many as yet unexplored ways. For the sea is a storehouse, not alone of dissolved minerals, but also of compounds that have great medical value. There are many health faddists who champion the consumption of seaweed, for example, which is rich in iodine, vitamins and other nutrients helpful to mankind. This sea plant is only one, and a most obvious one at that, that produces healthful compounds. Sea water always carries a high concentration of dissolved organic material produced by plants and animals. By analyzing sea water carefully, scientists have found small particles called metabolites. These are produced by some plants or animals and eaten by others. On analysis they have been found to contain not only vitamins, but also antibiotics—enzymes essential to the health of man—as well as many other medicinal forms of material. Intense research is now going on to isolate the compounds formed by that great chemical plant, the sea, and so far results indicate that we may actually find powerful drugs, anticancer medicines and tranquilizers. Some shellfish have fluids within their bodies that can help overcome infantile paralysis. The octopus secretes a powerful medicine that can be used on those suffering heart attacks.

There is one great healthful part of the ocean that man has just begun to attempt to get—the water itself, fresh and drinkable. No other natural material that is vital to man's existence is in such great demand today. Modern industry uses fantastic amounts of water to produce the materials we take for granted. In many cases, water is necessary to produce the product being used in one or more steps in manufacture as a cleansing agent, or solvent, for the materials. It then is discharged into the streams and rivers from which it was taken. The result is pollution of water on a scale we have never known before. The situation now is so critical that the Federal Government has begun a long-range campaign to purify and conserve water sources throughout the country. In addition to the growing use of water by an industrial civilization, two other factors are making water a scarce, and therefore valuable, commodity: one, the growing population requiring more water, and two, the many years of drought experienced in several parts of the United States.

At present over 500 million gallons of water are used by Americans every day of the year. Experts have estimated that by the year 1970 we will not have enough fresh water to meet this and the expected increased demand, unless new sources of water are found and tapped. Just off the shores of the world's continents lies 97% of the water on earth, and it is not drinkable in its natural state.

The problem of desalination is the one being tackled now not only by industry but by the U.S. Navy. The aim of the Navy is obvious. It must frequently maintain outposts far from conventional fresh water sources. The Navy experimental plants already are usable, although they are not economical so far as ordinary industrial use is concerned. Not long ago, when Fidel Castro threatened to cut off fresh water supplies to our base in Guantánamo, Cuba, the Navy was prepared to fly in a desalination plant and supply our forces there.

Industry is working to bring down the cost of desalting water to supply itself. The methods used are experimental and various in the extreme. One approach tries to evaporate the water economically (which leaves all impurities behind in the process), while another makes use of freezing salt water, which forces the salt and other chemicals to the surface where they can be removed.

So far all the methods tried have shown promise, but until more efficient plants are built, industrial water—which must be cheap—cannot come from the sea. Such methods do, however, provide water for human consumption at prices that are comparable now to those paid by city dwellers who must buy from commercial water companies.

The growing desire on the part of many countries to extract the riches of the sea has put an unusual strain on international relationships. The fact of the matter is that until recently, the sea, beyond a fringe near a country's shoreline, was considered open to all. Historically, there have rarely been attempts by some countries to claim more than a narrow strip of ocean along their shores as territorial waters. Russia, for example, tried to close the entire Bering Sea to vessels of other nations in the mid-nineteenth century in an attempt to prevent them from catching seals, but she was unable to make that prohibition stick.

Not many years ago the United States Government got into a quarrel with one of its own states—Louisiana—over who owned the offshore oil deposits found there. In a way it was a strange civil war of legalities. The reason for it was simple. If Louisiana owned the land, then oil drawn from it would be subject to the tax of that state and the benefit would go to the citizens of Louisiana. If the Federal Government owned it, the tax on oil taken from those waters would go to Washington. The case finally was resolved in the Supreme Court and the ruling gave the rights to the state of Louisiana. Other states, like Texas and California, benefited from that decision since their offshore oil deposits then represented a source of revenue to those governments. But, in the wide world, the situation cannot be decided by a Supreme Court whose decisions will be respected.

The problem has grown so acute that the United States has recently announced that its territorial waters reach not to the former 12-mile limit, but to the edge of the continental shelf. This extends the actual land area of the United States by ten per cent, a region of more than 360,-000 square miles that is open for development. Other nations, particularly smaller ones on the edge of the sea, are looking more enthusiastically at their ocean resources. Frequently they are backward nations possessing little in the way of technology or modern agriculture. Sometimes they are island nations and they are beginning to claim the edges of their lands long after they have been lost from sight beneath the waves.

This struggle for ownership of undersea lands will have increasing importance in future dealings with nations that are exploiting the sea. The Soviet Union is already one of the leaders among nations in exploiting the sea for food, profit and dominance. The problem has become so acute that a few leaders in the United States have begun to wonder whether military control under the sea is not more important to national survival than control of the surface of the moon. They point out that while getting to the moon is more dramatic so far as the public is concerned, the nation that controls the deeps, the ocean floors, and gains the greatest knowledge of the vast currents and turbulences, the nation that develops the most advanced weapons and undersea vessels may easily control the nation that lands on the moon.

For years the United States has been hampered in oceanographic research for the democratic reason that so many different interests were involved. The Navy, naturally, was interested. But the sea has a tremendous effect on weather, so meteorologists were also interested. The Department of Commerce was involved with shipping and with the fishing industry. The Coast Guard, the Coast and Geodetic Survey and a host of private institutions, colleges and industries all were pursuing their own ends independently of anyone else.

With the creation of the National Aeronautics and Space Agency there came the growing realization that a similar co-ordinating agency for oceanographic work should be created. Opponents of an over-all organization claimed that competition was a good thing. Ordinarily it is, but the taxpayer's dollars being spent through the government branches interested in the ocean were supporting the competition. Government grants and subsidies keep the competition going at a cost that private industry is unable to pay out of its own pocket. The last year has seen the creation of the first unifying organization for sea research in the nation's history. Congress has passed a bill creating a National Council on Marine Resources and Engineering Development.

This is a temporary organization, whose avowed aim is to give way at a later date to some other more permanent form of co-ordination among the more than 20 government agencies involved in sea research.

The aims of this agency, which is headed by the Vice President, are clearly stated, although they are so broad that nearly anything that has to do with the sea can be included under one of them. They are:

1. Acquiring the ability to predict and ultimately control phenomena affecting the safety and economy of seagoing activities.
2. Undertaking measures required for fullest exploitation of resources represented by, in and under the sea.
3. Employing the sea to enhance national security.
4. Pursuing scientific investigations for describing and understanding marine phenomena, processes and resources.

Veteran oceanographers view this list as a tall order to fill. In fact the scientists, politicians, engineers and interested businessmen who created the program made one tacit admission in setting up the objectives of the agency. As broad and complete as the agency seems to be, they admit the sea itself is broader and bigger—and deeper. The feeling among them was that the ocean was too big to be considered in its entirety for their purposes. They therefore proposed that the program concentrate on specific problems in specific areas where men could "conquer" the ocean a bit at a time. This was an interesting admission by men accustomed to thinking they both could and should conquer nature. Just possibly, some mythical god of the sea may have smiled, reading that proposed plan while tucked away in some grotto in a fathomless reach of the ocean depths.

8. SEA TECHNOLOGY

THE announcement "on station," became the litany aboard the *Chain*. As the ship proceeded southward, the weather changed, subtly at first, then suddenly one day work was carried on in what were almost tropic waters. The sea was sapphire blue in the strong sunlight. Bulkhead doors were locked open and the warm wind wandered through the labs. Off-duty scientists became sun worshipers, and some of them made serious efforts to gain a tan, lying on the foredeck reading, wearing sunglasses and occasionally turning to toast another side.

One Sunday the energetic cook announced a picnic, which seemed to be a strange occurrence in the trackless ocean. But he made the cookout a festive holiday affair.

A large oil drum, cut in half like a watermelon, became the cooker. Supported on welded iron legs and placed on the foredeck, it held glowing charcoal. On it the cook, decked out with a tall chef's hat and spotless white apron, turned steaks in a leisurely fashion. Potato salad and potato chips appeared. Olives, pickles, hot rolls, pie and ice cream were spread on a white-clothed table. Within minutes after the signal was given, everyone was munching the delicately done steak, buttering hot rolls or devouring olives and pickles.

The ship lay to in the tropic swells. Though no one had to contend with ants or mosquitoes, what was spread before them was quite obviously a picnic.

After the long trip southward this small interlude came as a total refreshment. But it was soon over and once again the *Chain* steamed toward her next station. The direction now was back along the carefully plotted transect, homeward bound. But on the way the *Chain*'s mission was to pick up those stations that were passed by in the stormy weather. The process of gathering data on this well-traveled profile of the sea probably will not be complete for several more years. It is a test section for the biologists and physical oceanographers of Woods Hole Oceanographic Institution, a continuing study started years ago and becoming, hopefully, one of the few truly "understood" places in the ocean.

After the picnic the usual routine soon settled again on

the ship's company. Daylight and darkness had no meaning. The question became merely: When do we get to the next station? Midnight. Three in the morning. Five A.M. Noon. It did not matter. The bridge called down the words, "on station," and the sleepy scientists tumbled out of bunks, hauled on socks, shirts, pants and hats and went on deck, yawning, thinking of the next dredge, the next plankton tow, the next hydrocast.

The weather once more turned brisk. Rain sifted down from an overcast sky. The water temperature dropped. In the Gulf Stream it rose nearly 20 degrees, which had been expected. The ship hove to on a station in the midst of that great flowing ocean river. The scientist in charge of the hydrocast duly put his instruments over the side but without much hope of success. He had tried earlier to get a profile of water with his Nansen bottles in this region, and so powerful was the current that his cable, weighted though it was, would not reach bottom. The force of the current drew the heavy cable out like a kite string and the scientist abandoned his effort in disgust. The same thing happened with the bottom dredge. The heavy dredge went overside and great lengths of cable were payed out. But the Gulf Stream took the dredge and held it horizontally within its wide, deep current. The station had to be bypassed.

North and west of the Gulf Stream the *Chain* began to attract sea gulls again. And out of the cloudy, rain-filled skies there came a land bird, a brown thrasher, driven to sea by the winds and lost in a strange environment. The wet, unhappy creature fluttered on deck until someone caught him in a net. He perched in a cabin until his feathers dried, took a few pecks of bread, drank some fresh water and then departed. He was the first messenger from the land that lay beyond the horizon.

One station remained, north of the Gulf Stream, and when all measurements there had been made, the rumble of the cable drum stilled. With the last specimen washed, examined and stored for further study in the laboratory, the *Chain* set a swift course homeward.

The senior scientist called all oceanographers to his cabin for a mild celebration. It was the senior scientist's treat and a tot of bourbon was measured out in cup and glass for all present. The talk turned to families, to the laboratory, to future projects and future voyages.

Then Gayhead Light lay abeam. The *Chain* neared Woods Hole, passing the long line of the Elizabeth Islands on her port side. The channel buoys leading to the harbor bobbed bright red in the gray, cold waters of Vineyard Sound and the *Chain* slowed, turning toward the dock.

Arrival was curiously like departure. The weather was gusty and cold. The sky was overcast. As the ship maneuvered cautiously toward the pier, the group of wives and children and friends waved mittened hands.

There was one difference in the landing. Another craft

was in view, one that had not been there when the *Chain* left. She was so different from anything else in the harbor that she could not escape notice. She was a small, white, undersea three-man submersible known as the *Alvin*. Like a realization of a Jules Verne dream, the vessel hung in midair, above the water, supported on her platform. Sleek and globular, she looked as strange and new as tomorrow.

The compact 26,000-pound mass of steel and fiberglass gives any onlooker dramatic evidence of advances in ocean research. Few people view the *Alvin* without thinking of the great nineteenth-century storyteller, Jules Verne, and his then improbable book: *Twenty Thousand Leagues under the Sea*. When that masterpiece was written, the world read in disbelief of the adventures of Captain Nemo. Today it still is a tale to capture the imagination as few other books do. But today we are on the thresheld of realizing Jules Verne's dream.

The *Alvin* is only one of many new, experimental research vessels that are taking men into the watery depths for extended periods of time. This small ship, built expressly for the scientists of Woods Hole Oceanographic Institution, can descend to depths of more than 1000 meters. It can explore all the sea floor along the continental shelf, staying submerged for as long as 24 hours and proceeding through the new environment at a speed of nearly seven miles per hour.

A larger, more versatile craft is under development now

for use by Woods Hole scientists. It is the *Aluminaut,* a three-man submersible that will be able to reach depths of 5000 meters and open the Atlantic Ocean basin to the direct observation of oceanographers for the first time.

Pioneers in this business of undersea exploration were William Beebe and Auguste Piccard. In 1934 Beebe made a record descent in a crude bathyscape which was nothing more than a steel sphere supported by a steel cable from a surface ship. Having no power plant of its own and no means of steering, the bathyscape was at the mercy of currents and thermoclines as it was lowered into the darkness beneath the sea.

This pioneer effort, however, attracted the attention of many scientists and one of them. Auguste Piccard, whose popular fame had come from his balloon ascents into the stratosphere, began to study the reverse problem—descent into the sea. The result finally came in 1960 when his son, Auguste Piccard, Jr., and a U.S. Navy Lieutenant, Donald Walsh, dived in *Trieste II,* a submersible that was the culmination of years of design effort on the part of the Piccards.

Trieste II is essentially an adaptation of an airborne balloon or blimp. It makes use of principles that are well understood by lighter-than-air craft. Buoyancy is the secret there, and such balloons use helium in a large bag to provide it. The other element of such a craft is the passenger-carrying gondola or compartment. Exactly the same idea is used in *Trieste II*. But the "balloon" is a metal cylinder filled with gasoline (which is lighter than water), and the gondola is a suspended steel sphere beneath it. For additional weight, tons of iron pellets are contained in the craft. On beginning a dive, gasoline is released from the *Trieste* and sea water is allowed to fill the empty space. This makes the vessel slightly heavier than the water around it and it sinks. The rate of descent can be as much as 50 feet a minute. As the sea floor is approached, the rate of descent is slowed by unloading some of the store of iron pellets, thus lightening the *Trieste*. A chain is dangled from beneath the gondola and when it touches the sea floor and, link by link, rests more of its length on the sediment, the net weight of the *Trieste* is reduced and its rate of descent is slowed further.

In January of 1960 the *Trieste II,* manned by Piccard and Walsh, began a deep dive that would have impressed Jules Verne. For hours the vessel sank through the black waters of the Pacific while the scientist and the Navy officer kept in contact with watchful colleagues on the surface by means of a special sonar telephone.

The *Trieste* was descending into the deepest known area of the sea, the Challenger Deep in the Marianas Trench in the western Pacific Ocean. At about 11,000 meters (35,800 feet) the *Trieste II* came to rest on the bottom of the world. Searchlights illuminated the ocean bottom that had not known such light since the oceans were formed. For 20 minutes the oceanographers observed this unexplored portion of the world. There was life at that unimaginable depth. A bottom fish was seen in the glare of the lamps, and a small red shrimp wiggled by.

The temperature of all water at the bottom of the ocean is nearly at the freezing point. Darkness is complete and never changes. On the steel spherical gondola of the *Trieste* there was fantastic pressure from the weight of almost seven miles of water overhead.

Record explorations like this one are made without tempting nature too much. The men stayed at the bottom of the world for the planned 20 minutes and then began the long ascent to the surface and into the sunlight again. Aside from mere record breaking, the main point of the descent was to test the equipment, to see if man had within his grasp the technical means of exploring the sea as he had always dreamed of doing. In all phases the journey was a success.

Since that time a great number of undersea vessels for exploration have been designed and built. Today more and more complex vessels are being planned, and the future will find even more sophisticated machinery built to take man beneath the sea.

One of the most impressive working research vessels in the world today is the Soviet submarine, *Severyanka.* This is a converted naval war vessel of modern design. The *Severyanka* can cruise through the major oceans of the world, studying currents and sea life at the leisure of the scientists aboard her. She is fitted with thick glass windows protected by sliding steel plates. When the Soviet oceanographers reach any desired location, they simply press a button and the steel plates slide to one side, revealing the oceanic scene through the glass windows to the scientists within the submarine. Seated comfortably at their desks in an undersea laboratory, the scientists can make any observations and measurements they want. The submarine can drift idly in undersea ocean currents, measuring their speed and direction. The ocean floor can be directly observed and all forms of sea life sketched or photographed through the windows.

The most intensive research in undersea vessels is presently going into smaller ones than the *Severyanka,* however. One of the types now in use is the Diving Saucer conceived by the popular pioneer of the sea, Jacques-Yves Cousteau. The saucer-shaped craft can go to depths of 1000 feet carrying two men, and it has made scores of successful descents and explorations.

Dr. Piccard himself has sold his *Trieste II* to the United States Navy for naval research into the depths. But this by no means puts Dr. Piccard out of business. A new undersea vessel, designed by him, is now in use. Called the *PX-15,* the new vessel brings into being one of Dr. Piccard's pet dreams. The vessel will have what Piccard calls "vertical stability," ability to hover at neutral buoyancy at any desired depth. It will remain submerged for long periods of time—as long as several weeks—and will be placed in a known current such as the Gulf Stream. There, using no power and moving through the water silently, it will simply float in that mid-oceanic river, drifting along at between 36 and 48 nautical miles per day. The vessel carries five scientists who observe all ocean life, make temperature and salinity measurements and determine light intensities. The scattering of light is noted and tape recordings are made of the sounds produced by fish and other marine creatures. This last task has heretofore been impossible to do in a submersible because most of them are under power. The noise of any propulsion engine cuts down on the receptivity of any recording device that is attempting to gather the language of undersea animals.

With the entrance of the Federal Government into all phases of its agencies' activities in oceanography, more and more money will go into the construction of such undersea craft. In addition, industry looking for minerals and oil deposits is spending millions every year in the race toward profits. The President's Science Advisory Committee has reported that funds should be doubled to investigate and exploit the oceans, and by 1971, $600 million a year should be spent for this.

Much of this money will go for new instrumentation. High on the list of items to be designed and purchased are undersea vehicles. Already research has shown that instead of using steel spheres as the compartment of undersea craft, we should be looking to another common material: glass. Compared with steel, glass offers more strength per pound of material used, and it has the curious capability of growing stronger as the pressure from sea water increases. Scientists now calculate that glass spheres would permit a submersible to withstand 4,320,000 pounds of pressure per square foot, far more than any weight of sea water could produce at the greatest known depths.

As may be expected, the means of atomic power for these small underwater vehicles is being built into them now. The Electric Boat Division of General Dynamics Corporation, the maker of our nuclear submarine fleet, is finishing a small submarine of approximately 60-foot length. This deep-diving craft is expected to take two scientists and a crew of five men to the ocean depths and have the capability of remaining there for several weeks for direct observation and the retrieval of specimens.

So great is the demand now for undersea exploration equipment that some companies developing submersibles lease them, complete with monitoring surface ship and expert skipper aboard the submersible, to take an interested party anywhere in the world to explore for science, for wealth or for historical relics lying on the floor of the sea.

Direct observation of the undersea world by men is necessary for complete understanding of that strange environment, but there are many measurements that can and are being made by instruments alone. In fact some measurements that men want are not observable without the instruments.

In the solid brick and concrete buildings of Woods Hole Oceanographic Institution, lying just a few steps away from the *Chain* at dockside, scientists and engineers have been working for many years on the development of such instruments.

The machine that made a continuous record of ocean depths on the *Chain*'s journey, for example, the precision graph recorder, was developed within the institution. Using the methods of acoustical echoing devices, the PGR, as it is called by officers and scientists alike, can perform a number of tasks that both assist navigation and bring data concerning the sea floor into the record books of the scientists.

To record the varying temperature of sea water as a research ship such as the *Chain* passes through a portion of the ocean, institution scientists have developed a trailing chain that will report temperatures electrically, on signal, going from the surface to a depth of 600 feet. There also is an electrically operated salinometer that will measure the amount of dissolved salt in sea water by determining its conductivity. A system of anchored buoys that can be put in place record automatically the current patterns over a large area of the sea. Aerial observations with the infrared surface temperature equipment give scientists a picture of temperature variation over a wide area in a very short time.

All such data, which lie primarily in the sphere of the physical oceanographer, are put into computers, either in the shore installation or aboard the large ships of Woods Hole, the *Chain* or the *Atlantis II,* one of the most modern oceanographic survey ships in the world.

As more and more questions about the sea are asked and as the data necessary to answer the questions are put through the computers and analyzed, the inevitable result is that still more questions come to mind and demand further research. It is this endless task in the realm of basic research that lies at the heart of the effort at Woods Hole.

It may seem strange that the *Chain* on its just completed voyage made use of techniques and instruments that have been in existence for many years. The Nansen bottle, for example, used in the hydrocast was developed early in this century by the great arctic explorer and oceanographer, Fridtjof Nansen. Ingenious, nearly foolproof and very sturdy, the Nansen bottle is an excellent example of the kind of instrument needed to withstand the forces of the sea. These forces—of waves and currents—have been well dramatized. What has not is the chemical "force" of its dissolved chemical compounds. They are extremely corrosive. When metal comes in contact with these salts there is nearly always fatal corrosion of the metal. If two dissimilar metals are used in an instrument and not protected from the sea water, an electric current can flow from one to the other through the medium of the salt water. It is, in effect, a battery. This current can increase the rate of metallic dissolution in the water as well as interfere with delicate electrical measurements.

The requirements, then, of any instruments immersed in the sea are sturdiness, resistance to corrosion and simplicity of design. The fewer parts there are, the fewer things can go wrong. And with all our ingenuity and electronic knowledge, these requirements must be met when it comes to measuring the sea.

The ultimate aim of our technological effort to probe the sea seems at first glance to be unattainable and incredible, as far removed from reality now as it was when Jules Verne wrote of it: to establish a base or laboratory on the ocean floor. Men, it is believed, will live there for extended periods of time. They will come and go as they please, rising to the surface or dropping to the depths for another "tour of duty" in some sea colony.

Yet, in cautious steps, men are doing just that today. Under a U.S. Navy program, large metal structures have been built to house men living under water. The first, called Sealab I, was anchored on the bottom off the coast of Bermuda in nearly 200 feet of water. Navy divers lived and worked in that chamber for more than ten days. Sealab II has been anchored at a similar depth off the coast of California and oceanographers and technicians—ten in all

—have occupied that colony on the sea floor. It is true that the depth is not impressive, but the problems at 200 feet are critical enough for the present.

The strange laboratory is supplied by lines connecting it to a hovering surface vessel and to the shore installation not too far away. The men who go down into these first colonies take special training. Their bodies will be subjected to air pressure as great as that of the surrounding water pressure. They put on scuba gear and step into the water from the compressed atmosphere of a compartment of the lab that is open to the water on the bottom. The divers move freely in the ocean, carrying their breathing apparatus or sometimes without using it, but using a long hose to which they go frequently to replenish their supply of oxygen.

In Jules Verne's terms these crude attempts are not very dramatic. The labs are close to land, are constantly watched over by men in surface vessels, and must get their supplies of power and fresh water from above. But the day is coming when more advanced undersea laboratories will be built.

Nuclear power, of course, offers the answer to a source of energy for the operation of instruments and a supply of electricity. But the question of air, or rather of oxygen, remains.

To solve this problem, the Naval Civil Engineering Laboratory is conducting interesting and encouraging experiments. Its scientists are making use of sea water itself to help them. It is known that even at great depths sea water contains dissolved oxygen. If it did not, no life could exist in it. There are some known "dead spots" in the ocean where the lack of dissolved oxygen prohibits the growth of any form of life. But most of the ocean has some oxygen.

Sea water has another helpful advantage. It will absorb carbon dioxide readily. Most sea water has little free carbon dioxide in it. In other words, it is relatively rich in oxygen and deficient in carbon dioxide. Stale air, exhaled by lungs, has little oxygen and a great deal of carbon dioxide.

As long ago as 1953 sea water was used to withdraw carbon dioxide from stale air in submarines. Now the problem faces the Navy experimenters of exchanging carbon dioxide for the dissolved oxygen in the ocean water.

To do this they had to have a special membrane that would permit the passage of gases but not liquids. Fortunately a company in the United States developed a usable silicone rubber material that would do the job.

Using this membrane, an underwater chamber has been built and laboratory animals placed in it. When sea water is circulated along the outside surface, the carbon dioxide in the exhalations of the animals passes out through the membrane wall. Oxygen from the sea passes through toward the inside atmosphere. In controlled experiments, the animals have lived for as long as two weeks in such a chamber.

But since sea water contains only one-thirtieth the oxygen that air does for equal volumes, a great deal of sea water has to contact the membrane to supply the animals living within the chamber. Another disadvantage lies in the speed with which the oxygen diffuses or passes through the membrane, and the carbon dioxide is passed to the outside sea water.

Experiments proceeded in an attempt to increase the rate of exchange of both gases. In the most successful one yet tried, a direct mingling of air bubbles with sea water produced encouraging results. The air that is to be "scrubbed" or reoxygenated is bubbled through a chamber through which sea water is circulated. In this situation, the gas bubbles are moving upward through the water and there is no separating membrane of any sort. The exchange of oxygen and carbon dioxide was discovered to be almost instantaneous. As soon as a bubble entered the mixing chamber, its carbon dioxide content dropped and the oxygen content rose sharply.

To test the apparatus, a small burner that consumed only oxygen was placed in the air chamber that was connected to the exchange container. Tests, run without exchanging air, showed the burner would operate for 31 minutes before a lack of oxygen put it out. With the exchanger operating, the burner remained in operation for more than four hours, and then was finally put out by the scientists. It would have burned indefinitely. A next step was to introduce living animals. Two small rats were placed in the "habitation chamber," as it is called, and lived comfortably for eight hours. On being removed their health was checked and found to be perfect. Two weeks later, no ill effects could be detected.

The aim of men in conducting such experiments is clear: preparation for the establishment of undersea colonies, permanently built on the sea floor and independent of air or fuel supply from the land. The meaning of this aim is truly astonishing. Men may some day live and work at the bottom of the sea for weeks and months at a time.

Both the Navy and industry are understandably excited by the prospect. Scientists, of course, will be able to examine the sea floor and the life in sea water at their leisure. The Navy's interest is defense. Hidden bases on the sea floor and the development of submarine bases far below the surface of the sea are just two of the many possible uses of such a development.

For industry the prospect is of great interest. With an undersea base built, for example, on a vast field of manganese nodules or a huge phosphate deposit or, more obviously, an oil field, men can extract the wealth of the sea in the least costly and least hazardous manner: by direct mining and manipulation of the equipment from within a great sheltering dome just as they would on the earth's surface.

9. FAMOUS OCEANOGRAPHERS AND VOYAGES

CRUISE 58 of the *Chain* was over. By ordinary standards it was not a long voyage nor did it produce startling discoveries to catch the interest of the scientific world. What it did was add just a bit more to the growing body of knowledge about the sea that men have been gathering for years.

At times on the cruise just completed, the scientists would sit in the ship's comfortable and well-stocked library and their talk, in moments of leisure, frequently turned to an earlier ship of Woods Hole Oceanographic Institution, a ship many of them had sailed on, the *Atlantis I*. The reminiscences of the men betrayed at times a thoroughly romantic interest in ships and in the sea. For the *Atlantis I* was a sailing vessel and she was the first oceanic research vessel designed and built for the task of oceanography in the United States.

Ketch-rigged, the 142-foot auxiliary-powered vessel was constructed in Denmark and completed in the year that Woods Hole Oceanographic Institution itself was founded, 1930. The first director of the institution was Dr. Henry Bigelow. To sail the *Atlantis I* across from Denmark to Woods Hole he chose a young and enthusiastic scientist and sailor, Columbus Iselin. The maiden voyage of *Atlantis I* was the first of many in the Atlantic, the Gulf of Mexico, the Gulf of Maine, the Mediterranean and in the South Pacific off the western coast of South America.

For most of the year, each year, for decades, the *Atlantis I* winged her way across the seas, far from shipping lanes, sampling the waters of the world. Scientists aboard the *Chain* had sailed on her and they recounted their true sailing adventures with enthusiasm. The ketch carried a crew of seventeen and had space for nine oceanographers. In the days of that ship, both scientists and crew stood watch, turn about. For, like all sailing ships, *Atlantis I* required the constant concern of everyone aboard her.

Not, of course, that all hands were equally competent.

The men of the crew were trained in sail but many of the oceanographers were little more than willing and enthusiastic amateurs when it came to handling a sailing ship. Several scientists aboard the *Chain* admitted ruefully that when it came their turn to stand watch, the track chart of the ship, which was kept to show the actual course of the vessel, began to look like the meanderings of a punch-drunk fighter. Where the skilled helmsmen kept the sailing vessel on a straight, clean course, the scientists' trick at the wheel frequently gave odd zigs and zags to the log. Handling a large sailing vessel in varying winds is a difficult art.

Throughout its years of service, *Atlantis I* contributed important information to the growing field of oceanography. Many pieces of modern equipment now standard on oceanographic research vessels were tested initially aboard the small sailing ship. The field of her most intensive work was the North Atlantic Ocean and the sweeping Gulf Stream that dominates the western portion of that ocean basin. Most of the research during the years from 1931 to 1959 concentrated on the physics and chemistry of the oceans. It was in these fields that the ship made her most important contributions.

The days of *Atlantis I* are over so far as the scientists of Woods Hole are concerned, but old ships somehow manage to survive. Today, refitted and renamed, the *Atlantis I* is in the service of the Argentine Government. Manned by another crew, the ship sails the waters of the South Atlantic, still pursuing the mysteries of the seas.

Oceanographers the world around generally are agreed that modern oceanography began with one spectacular cruise, that of the H.M.S. *Challenger.* It is sometimes called the Nares Expedition, in honor of its initial commander, Captain George S. Nares. But most oceanographic cruises still refer to the ship as the identifying name.

There had been earlier work of some importance in oceanography but in most cases the expeditions were interested in other aspects of the sea: charting shoals, identifying new lands or attempting to learn more about the earth's magnetic field to aid in marine navigation. Only incidentally did they examine the sea itself.

Among the earliest of those combination voyages were those of James Cook in the eighteenth century, the Russian Admiral Bellingshausen and the American Charles Wilkes in the first half of the nineteenth century. These were all exploratory voyages in which the principal aim was the discovery of new geography that could be claimed by the interested countries or the extension of a nation's influence in newly found lands for commercial gain and status.

The *Challenger* expedition, on the other hand, was designed for one purpose alone: the examination of the sea. The English had been examining the sea mysteries close to their shores for some time, and many questions came from the work they were doing. A Professor W. B. Carpenter suggested to the British Admiralty that a vessel be fitted out for the express purpose of oceanographic research and sent on a long cruise around the world. The English Royal Society, a scientific organization, was asked to draw up a comprehensive plan. When that was done, an English navy warship, the H.M.S. *Challenger,* was selected to carry the scientists on their long cruise. The ship was a full-rigged sailing vessel that had an auxiliary steam engine. Termed by the British a corvette, she ordinarily carried 18 cannon. All but one or two of these were taken off and the gundeck was turned into laboratories for the scientists.

In addition to the navy crew, six scientists were to board the ship on its departure from England. They were under the leadership of Wyville Thomson, who was a professor of natural history at the University of Edinburgh. He chose two other members of the university staff: John Murray and J. Y. Buchanon. From outside the university he selected J. J. Wile and Dr. Rudolph von Willemos-Suhm.

Equipment for the voyage included everything available at that time—the usual water sampling bottles, dredges, nets, trawls and untold miles of sounding wire and line to reach farther into the sea than any previous soundings.

For the navy crew the voyage was merely one more of a war vessel in those times when Britain in fact ruled the waves. Few if any of the officers or men of the *Challenger* realized that on this cruise they would sail, quite literally, into history.

On December 7, 1872, the *Challenger* stood out of the Thames for the open sea. At the time of her sailing, scientists were almost totally ignorant of the world seas. What information they had came from a few isolated soundings of the sea bottom in relatively well-traveled areas; wild theories about life in the sea, maximum depths, currents and water circulation were totally unsupported by experimental data.

When the H.M.S. *Challenger* returned on May 24, 1876, three and one half years later, oceanography as a science had been established. The six scientists had accomplished an amazing amount of work. The expedition had sailed around the world, covering waters in the North Atlantic, South Atlantic, the Antarctic, Indian and Pacific oceans. The ship's log showed the *Challenger* had sailed nearly 70,000 miles. In the process the scientists had taken specimens of water and sea life from more than 360 stations. Eventually the specimens revealed that the men had discovered over 4700 new species of plant and animal life and over 700 new genera. The depth soundings gave, for the first time in history, the outline of the topography of the ocean floor throughout the world. The principal current systems of the oceans were discovered and the water temperature regions fixed. For although the six scientists were principally interested in marine biology, they acted on that historic voyage as physical oceanographers as well. Weather data was taken on the trip and magnetic readings were gathered also to complete the monumental journey.

On returning to England, Wyville Thomson began the long task of analyzing the data they had gathered, for while the men were on board ship, there had been little time for more than a hasty approximation of what their work truly meant. Thomson called in experts from many countries working in different areas of science to help with the analysis and publication of the results. Today that record stands as the foundation stone of oceanography. Fifty volumes were published and though Thomson did not live to see the work finished, his colleague John—later Sir John—Murray took over and produced the work.

The accomplishments of the expedition are all the more astonishing when the crudity of the apparatus and the lack of modern techniques is recognized. Merely to take a deep sea sounding, for example, required many hours on station. And although a steel wire was carried to make soundings easier, it was found that good hemp line could do the job in less time. The *Challenger's* steam auxiliary engine was used to lower the bottom-sounding lead or dredge and to keep the broad-beamed heavily rolling ship on station through the hours of lowering a line. The men of the

Challenger worked in deep water, some of the deepest yet discovered, and their deepest sounding reached nearly 27,000 feet into the Marianas Trench in the Pacific. It was a record sounding for a record voyage.

* * *

If the *Challenger* expedition was a single monumental voyage of discovery, one man, Matthew Fontaine Maury, was a singlehanded monument in his own right. As a young United States naval officer, this man more than any other in oceanographic history contributed to the physical study of the ocean.

Maury was born in 1806 in Virginia and became a midshipman in the United States Navy in 1825. His first cruise took him around the world in the U.S.S. *Vincennes.* His regular duties as an officer in the United States Navy were restricted in 1839 when he was injured in an accident and permanently lamed. While Maury was essentially a sailor with a love of the active life, he did not allow his handicap to stop him in his pursuit of a career. It is fortunate for the science of physical oceanography that he turned his disability to advantage. In 1841 Maury was placed in charge of the depot of charts and instruments. This department later developed into the U.S. Naval Observatory and the Hydrographic Office where bathymetric charts are produced. It was here that Maury began his true life's work.

As a sailor he knew of the problems of navigating in strange waters where no man had knowledge of winds and surface currents that affected ships. These items of information were particularly important to ships operating near dangerous shoals. Maury's aim was to provide seamen with charts of all the oceans of the world showing the season-by-season directions and speeds of winds and currents. When he first conceived of studying the ocean currents and winds, he had very little information to work with. But as early as 1836 he had gathered enough data to publish the *New Theoretical and Practical Treatise on Navigation*. Three years later while still a subordinate in the depot of charts and instruments he conceived a plan to serve his aim in charting the oceans. He devised special logbooks for ship navigators and persuaded them in their travels to keep them up-to-date. But while waiting for this information to come to him, he studied old logbooks that were on file in the department. His first charts began coming off the presses within a few years. They were an immediate success. He sent them to ship navigators who found that by using the information Maury had given them they could cut sailing times drastically. Twenty, thirty, fifty, even a hundred and fifty days could be lopped from long voyages. And in the competition in trade during those years, such a saving of time meant increased profits for the owners. As navigators began to appreciate Maury's work, they were more prompt and more careful about recording the information he wanted, and hundreds of navigators sent in their logbooks.

From this great quantity of data, Maury wrote the first textbook in physical oceanography: *The Physical Geography of the Sea.* In it was information on climate over the vast reaches of the oceans, prevailing winds, currents and even the location of the best whaling grounds. Singlehandedly, Maury launched the still-continuing study of the ocean's physical features. His work showed the necessity of international co-operation in covering the seas. The interest he aroused among oceanographers and sailors of all nations produced one of the first international conferences on the subject in Brussels in 1853.

From the voyage of the *Challenger* in 1873, until the outbreak of World War I, study of the sea was carried on largely by individuals or at least individual organizations. In many cases the expeditions were primarily interested in land exploration and discovery, particularly in the polar regions. Most of the geography in low latitudes had been at least discovered; no new continents or islands of great size remained to be claimed for an exploring nation. But man had yet to reach the poles of the earth. Both arctic and antarctic adventurers went forth, some on private expeditions and others backed by geographic societies or, in a few cases, governments. In the process of trying to reach the poles, the scientists on the ships that sailed into ice-covered waters took specimens and soundings, charted even more currents and measured wind speeds and directions. Such work was standard procedure for the men of such expeditions. But of all such voyages in the last two decades of the nineteenth century, none was a purely oceanographic expedition. With one exception: the *Fram* under the leadership of the greatest polar explorer and scientist of all time: Fridtjof Nansen.

Nansen was born in Froen, Norway, in 1861, and from his early youth was fascinated by the mysteries of the arctic region, its ice-strewn seas and the barren desolate shores that rim the polar basin.

He determined at a very young age what he wanted to do in life: explore the northern oceans and lands. To prepare himself for this goal he entered Christiania University in 1880 where he studied zoology and biology. In March 1882 Nansen got his first taste of arctic existence when he joined a sealing ship that sailed to Greenland waters. More years of study followed until in 1887 he received his doctor's degree and set about preparing for the first great journey of what was to be a long and incredibly active life. The great ice sheet that covers Greenland had never before been explored far from the coastline. Nansen determined to strike out across that forbidding mass of ice with several companions to see exactly what the icecap was like. They made the crossing in 1888, reaching a height of more than 8900 feet, and successfully ended the journey on the opposite coast.

As his powers matured and his knowledge of the Arctic increased, Nansen allowed himself to dream of another more difficult journey, the one for which he is most remembered. It was Nansen's belief that through the Arctic Ocean there was general drift of water or possibly a current flowing on or near the surface that began north of Bering Strait between Alaska and Siberia and moved over the top of the world to discharge itself into the North Atlantic between Greenland and Spitzbergen. On what did he base his theory? He had collected, in his travels, bits of driftwood on the shores of Greenland that came from trees known to grow only in Siberia half a world away. He had also studied the strange case of the wrecked ship, *Jeannette*. She had been broken up by ice near the coast of Siberia above the Bering Strait. Three years after that unfortunate ship foundered, bits of wreckage and gear were found embedded in ice that made its way to the southern shores of Greenland.

Nansen knew as much as anyone at the time about the Arctic Ocean: very little. But he set out to prove his reasoning and produced a record oceanographic voyage that has never been surpassed. His scheme was to build a ship especially fitted to withstand ice pressure, deliberately sail that ship north through the Bering Strait and on north past the farthest land until she was imprisoned in the arctic ice. Then he and his crew would settle down for a three-year drift across the top of the world. When he first made the plan known to his friends and to the public, Nansen was regarded as a lunatic. Veteran explorers called his plan preposterous, but Nansen went ahead despite all the opposition. Nansen's famous vessel was designed by the internationally known shipbuilder Colin Archer, and it was unique among ships in that its strengthened wooden hull was exceptionally broad-beamed, with her underside rounded so that when the ice pressed horizontally against her sides she could simply rise and ride on top of the ice. Instead of fighting the immense pressures of polar pack ice, the *Fram* was going to let those pressures raise her from the water where she could rest comfortably on the ice itself.

In September of 1893 the voyage began as Nansen headed the *Fram* into the ice pack above Siberia. Although Nansen believed the voyage would take only two or three years, he had supplies for six years of isolation from civilization.

During the long drift that lay ahead, Nansen planned to cut through the ice, take water and bottom samples and generally run the ship just as a permanent shore-based laboratory was operated, except for the fact that his laboratory was moving on the ice as it drifted across the unknown sea beneath.

Data were gathered each day and regular recordings made of everything from wind velocity and direction to the salinity and temperature of water at varying depths beneath the ship. Specimens of arctic sea life came up in dredges and nets. Long hours were spent in analyzing the material that was gathered as the weeks and months passed.

At the end of the first year, the drift theory of Nansen did not appear to be working. The ice in which the *Fram* was imprisoned moved north and then south, jockeying a bit every day but not producing much progress toward the goal of drifting across the North Pole and out into the Atlantic Ocean. The geographic North Pole was one of Nansen's goals also. No one had yet reached it and he thought the ice drift could put him and his base ship within reach of that long-sought geographical prize.

As the second year began, the drift increased, moving the *Fram* north and east. It became evident to the men, however, that the *Fram* would not drift to the top of the world, although a polar current did indeed exist. It was going to take the ship several hundred miles south of the North Pole. By the middle of December in 1894, the *Fram* was at latitude 82° 30′ north, well over 400 miles from the pole.

The direction of drift indicated they would get no closer, and Nansen revealed a plan he had conceived as the drift deflected the ship from a northward course. He proposed to take one man with him and set off with dog sleds in an attempt to reach the pole across the ice pack.

He knew the risk was great but he was confident that they could survive. He also knew that he would never be able to find the ship once he had left her, for she would go on drifting in some erratic manner as the current dictated. He and his companion would be on moving ice also. Nansen therefore decided to head for the pole and then travel south across the ice to land. The entire crew remaining behind under the command of Nansen's trusted colleague, Otto Sverdrup, would continue to take data from the polar sea.

For days Nansen and his chosen companion traveled northward across the ice, only to find to their dismay that the southward drift in that vicinity was denying them much progress poleward. It was like marching on a treadmill. The two men did however reach 86° 14′ north, the farthest north of any explorer at that time. The two men finally turned back and headed for Franz Josef Land, which they reached on August 26, 1895. Food shortages forced them to winter over on one of the many islands making up that body of land, and on May 19, 1896, they started south toward Spitzbergen.

On the way one of the strangest meetings in arctic history occurred. Nansen and his companion, Johansen, sighted another exploring party of the Jackson-Harmsworth Expedition. They returned with this party aboard their ship the *Windward,* and reached Norway on August 13. One week later, the *Fram* appeared. She had completed her drift, been released from the ice just as Nansen had predicted, and all hands were safely home.

In his classic account of this last great oceanographic voyage, *Farthest North,* Nansen displayed all that had been learned about the physics of the Arctic Ocean basin. The knowledge he brought back rivals that of any expedition, if not in the breadth and scope of data, at least in the retrieval of information from the least hospitable sea in the world.

Within 20 years of the completion of the *Fram* expedition, the entire era of individual exploration had come to an end. The science of oceanography had grown to such an extent that nations with interests involving the sea began to allot money for its exploration and exploitation. In the same manner the era of outstanding individuals came to a close. Teams of scientists worked under national supervision. As the data poured in from scientists around the world, an increasingly accurate picture of the oceans began to emerge. Systematic surveys of the seven seas became the rule, and many ships moved from point to point along a line much like the one laid down for the *Chain* from Gayhead to Bermuda to gather data at regular check points.

World War II interrupted the growing cooperation among nations in the exchange of information on oceanography, but there was a hidden benefit in the interruption. The demands of war produced new and useful instruments for measurement and detection. Since the nations in the struggle were on a war economy, cost of developing these instruments was no object, and the new technology in which we find ourselves today was ushered in during those war years.

As the world settled back in the postwar years, oceanographers took up their work again but with many new tools to help them. Lines of communication opened once again, international scientific cooperation grew and culminated in the famed International Geophysical Year (technically spanning the time between 1957 and 1959). In this brief period, teams of scientists made similar measurements all over the earth, and in the oceans as well. So much new data was gathered that centers of information became necessary.

As might be expected, the areas of the world ocean that have received the most attention are those adjoining the more highly developed countries. The North Atlantic is probably the most studied ocean in the world. Next in interest is the South Atlantic. Recent joint studies by several nations have been made in the Indian Ocean while the Arctic has been penetrated and sampled by nuclear submarines as well as surface vessels. With our interest in Antarctica as a continent for scientific research, the seas of the Southern Ocean surrounding that frigid mass of ice and rock have been penetrated and measured. Part of the interest in those inhospitable waters comes from the fact that hunting for whales has taken men into those high southern latitudes for many years. A more modern reason for investigating antarctic waters is that the Antarctic Continent and its surrounding sea actually create the weather situation that affects the entire Southern Hemisphere. The least studied ocean remaining is the South Pacific, and indeed so little was known of it at the outbreak of World War II that the U.S. Navy had to rely on charts laid down over one hundred years ago by the United States Exploring Expedition under the command of Lieutenant Charles Wilkes. Part of the reason for ignorance concerning the South Pacific is that it is in one of the least developed areas of the world, with the exception of that portion of it near New Zealand and Australia. But another factor is its size. The broad Pacific from the Aleutian Islands south to the Antarctic presents a formidable body of water to analyze and understand.

As more and more money is earmarked by countries for oceanographic research, however, this last unknown region of the world ocean will eventually yield to systematic, technical attack. The United States alone is building research vessels at the rate of five a year. The Soviets already have an impressive fleet of the largest research vessels of any nation on earth. Britain, Norway, Japan and others are also adding to their fleets. The number of shore-based installations like Woods Hole, Massachusetts, are growing rapidly. In only a few years, the science of oceanography will have developed the network of ships and stations plus data processing and retrieval stations to place the oceans of the world under constant surveillance.

On the following pages, a list of the outstanding oceanographic voyages shows the extent of early oceanographic research and the important work done by the ships of many nations. The material was gathered by Professor Georg Wüst and published in an article titled *Deep-Sea Expeditions and Research Vessels, 1873–1960*. The author is indebted to Pergamon Press Limited, Oxford, England, for permission to include this valuable survey. It appeared in *Progress in Oceanography,* Volume 2.

REPRESENTATIVE VESSELS IN OCEANOGRAPHIC DEEP-SEA RESEARCH (1873–1960)

Name of Vessel	Nationality	Commissioned	Ocean	Main Studies
		Ia. Era of Exploration (1873–1914)		
Challenger	British	1873–1876	Atlantic, Indian, Pacific	Biology, Physics, Sediments
Gazelle	German	1874–1876	Atlantic, Indian, Pacific	Physics
Blake	U.S.A.	1877–1886	Atlantic (Gulf Stream System)	Physics, Currents
Albatross	U.S.A.	1887–1888	North Atlantic	Biology, Physics
National	German	1889	North Atlantic	Plankton
Fram	Norwegian	1893–1896	North Polar Sea	Physics
Valdivia	German	1898–1899	Atlantic, Indian	Biology of great depths

REPRESENTATIVE VESSELS IN OCEANOGRAPHIC DEEP-SEA RESEARCH (1873–1960) (cont'd)

Name of Vessel	Nationality	Commissioned	Ocean	Main Studies
Princess Alice I and II Hirondelle I and II	Monaco	1888–1922	North Atlantic	Biology
Gauss	German	1901–1903	Atlantic, Indian, Antarctic	Physics, Geography
Planet	German	1906–1907	Atlantic, Indian, West Pacific	Physics
Deutschland	German	1911–1912	Atlantic, Antarctic	Physics
Ib. Transition to systematic research (1904–1924)				
Michael Sars	Norwegian	1904–1913	Norwegian Sea, North Atlantic	Physics
Armauer Hansen	Norwegian	Since 1913	Norwegian Sea, North Atlantic	Physics

II. Era of National Systematic Surveys (1925–1940)

Meteor	German	1925–1927	German Atlantic Exp. 65°S–20°N	Physics, Chemistry Meteorology, Geology, and Biology
Meteor	German	1929–1938	Iceland-Greenland waters, some N. Atl. cruises	Physics, Chemistry Biology
Dana I and II	Denmark	1921–1935	Atlantic, Indian, Pacific	Biology
Carnegie	U.S.A.	1928–1929	Pacific	Physics, Biology, Sediments
Willebrord Snellus	Dutch	1929–1930	Indonesian Seas	Physics, Sediments
Discovery and Discovery II	British	1930 ff	Antarctic	Physics, Biology
Atlantis	U.S.A.	1931 ff	North Atlantic	Physics, Chemistry

REPRESENTATIVE VESSELS IN OCEANOGRAPHIC DEEP-SEA RESEARCH (1873–1960) (cont'd)

Name of Vessel	Nationality	Commissioned	Ocean	Main Studies
Ryofu Maru	Japanese	1937 ff	Pacific	Physics
E. W. Scripps	U.S.A.	1938 ff	Pacific	Physics, Chemistry Sediments
Altair	German	1938	Gulf Stream Exped. N. Atlantic	Physics, Currents
Sedov	Russian	1938–1939	North Polar Sea	Physics
IIIa. Period of New Geological, Geophysical, Biological and Oceanographical Methods (1947–1956)				
Albatross	Sweden	1947–1948	Atlantic Indian	Sediments, Physics
Galathea	Denmark	1950–1952	Pacific	Biology of abyssal depths
Atlantis	U.S.A.	1947 ff	Atlantic	Physics, Geophysics
Vema	U.S.A.	1953 ff	Atlantic, Indian, Pacific	Geophysics, Geology

Anton Dohrn	German	1955–1956	North Atlantic	Biology, Physics
Spencer F. Baird	U.S.A.	1956	Pacific (operation Chinook)	Geophysics, Physics
	IIIb. Transition to synoptic research in smaller areas (1950)			
Atlantis and other vessels	U.S.A.	1950 (1956)	Gulf Stream	Physics, Dynamics, (Operation Cabot and second survey)
	IV. Era of International Research Cooperation since 1957			
Crawford	U.S.A.	1957–1959	Atlantic IGY Program	Physics, Chemistry
Atlantis	U.S.A.	1957–1959	Atlantic IGY Program	Physics, Chemistry
Discovery II	British	1957–1959	Atlantic IGY Program	Physics, Chemistry
Capitan Canepa	Argentina	1957–1960	Atlantic IGY Program	Physics

REPRESENTATIVE VESSELS IN OCEANOGRAPHIC DEEP-SEA RESEARCH (1873–1960) (cont'd)

Name of Vessel	Nationality	Commissioned	Ocean	Main Studies
Gauss	W. Germany	1957–1960	North Atlantic Polar Front and Overflow Program	Physics
Anton Dohrn	W. Germany	1957–1960	North Atlantic Polar Front and Overflow Program	Biology, Physics
Explorer and other vessels	Scottish	1957–1960	North Atlantic Polar Front and Overflow Program	Biology, Physics

Vitiaz	U.S.S.R.	1957–1960	All Seas (IGY Program)	Physics, Chemistry, Biology, and Geology Trenches
Numerous research vessels	Japan, U.S.A. Canada	1955–1960	Pacific (Norpac Program)	Physics, Biology
Mikhail Lomonsov	U.S.S.R.	1957–1960	All Seas (IGY Program)	Physics, Chemistry, Biology
Ob	U.S.S.R.	1957–1960	Southern Seas (IGY Program)	Physics, Chemistry, Biology
Horizon	U.S.A.	1957–1960	Pacific	Geophysics, Physics
Chain	U.S.A.	1958–1960	Atlantic, Indian	Geophysics, Physics
Argo	U.S.A.	1959–1960	Pacific, Indian	Geophysics, Physics
Shokalski	U.S.S.R.	1960	All Seas	Physics, Meteorology
Woeikof	U.S.S.R.	1959	All Seas	Physics, Meteorology
Pole	U.S.S.R.	1959	—	Physics etc.

REPRESENTATIVE VESSELS IN OCEANOGRAPHIC DEEP-SEA RESEARCH (1873–1960) (cont'd)

Name of Vessel	Nationality	Commissioned	Ocean	Main Studies
Lena	U.S.S.R.	1958	Polar Seas	Physics, Meteorology
Severyanka	U.S.S.R.	1958	All Seas	Submarine for oceanographic research
Numerous research vessels	International	Since 1958	Indian Ocean Expedition Program	Physics, Geophysics Biology, Chemistry, Geology, Meteorology

INDEX

Index

Index

Index

Index

Index

DEC 8 1967

PAID FEB 28 1981

SOUTH BR.
TEEN TITLE

551.46 BIXBY, WILLIAM
Seawatchers.

3

FORT WORTH PUBLIC LIBRARY

PLEASE DO NOT REMOVE SLIP OR PLASTIC CARD FROM THIS BOOK.

If this book is returned to the library without plastic book card or slip, there will be a minimum charge of 25¢ each.